M000274706

FINANCIAL ASTROLOGY ALMANAC 2021

Trading & Investing Using the Planets

M.G. Bucholtz, B.Sc, MBA, M.Sc.

Wood Dragon Books

Financial Astrology Almanac 2021 – Trading & Investing Using the Planets

Copyright © 2020 by M.G. Bucholtz.

All rights reserved. No part of this book may be used or reproduced in any manner whatsoever without written permission except in the case of brief quotations embodied in critical articles or reviews.

Published by :
Wood Dragon Books,
Box 429, Mossbank, Saskatchewan, Canada, S0H 3G0
http://www.wooddragonbooks.com

ISBN: 978-1-989078-44-0

For information contact the author at: supercyclereport@gmail.com

DEDICATION

To the many traders and investors who at some visceral level suspect there is more to the financial market system than P/E ratios and analyst recommendations.

You are correct. There is more. Much more. The markets are rooted in astronomical and astrological timing. This book will add a whole new dimension to your market activities.

DISCLAIMER

All material provided herein is based on material gleaned from mathematical and astrological publications researched by the author to supplement his own trading. This publication is written with sincere intent for those who actively trade and invest in the financial markets and who are looking to incorporate astrological phenomena and esoteric math into their market activity. While the material presented herein has proven reliable to the author in his personal trading and investing activity, there is no guarantee this material will continue to be reliable into the future.

The author and publisher assume no liability whatsoever for any investment or trading decisions made by readers of this book. The reader alone is responsible for all trading and investment outcomes and is further advised not to exceed his or her risk tolerances when trading or investing on the financial markets.

Contents

INTRODUCTION

Many traders and investors think company press releases, media news opinions, quarterly earnings reports, and analyst targets drive stock prices and major index movements.

I disagree. I believe the financial markets are defined by various cycles of planetary activity. Overlap and string these cycles together over time and you will have the ups and downs that characterize a stock chart, a commodity price chart, or the chart of a major index.

Cycles of planetary activity include the Jupiter/Saturn Gann Master cycle which unfolds over two decades, the times between events of Mercury or Venus being retrograde, and the times between Mercury and Venus being at elongation extremes. Cycles also involve the 18.6 year movement of the North Node through the signs of the zodiac, and the declination of planets above and below the ecliptic plane. At the shorter end of the spectrum, the cycle from one new Moon to the next can also be seen to have a bearing on the market. There are even cycles tied to the Hebrew calendar and to the Kabballah.

Even though the mainstream media refuses to embrace astrology or cycles of any sort as valid tools for timing the markets, there are powerful players in the major financial centres of the globe who do embrace astrology and its cycles. They use these cycles to their advantage to make money in an uptrending market. They also use these cycles to induce trend changes across markets. As the trend turns and markets start to fall, these players profit from their short positions while the average investor on the street experiences angst and sleepless nights knowing the markets are trending down and against them.

"How and why these various cycles have come to be?", is the burning question that remains unresolved. As I have studied these cycles, I have developed a new sense of awe for what I deem to be a higher power that guides the Universe. After reading this Almanac, you too may have reason to pause and ponder the power of the cosmos.

"Who are the powerful players are that use astrology to move markets?", is another burning question. Is it a select group at J.P. Morgan? Is it a group in a dark-panelled office in London? I will likely never know. But what I do know is, that human beings paying attention to planetary cycles is not a new concept. Ancient civilizations as far back as the Babylonians recognized cyclical activity, but in a more rudimentary form. Their high priests tracked and recorded changes in the emotions of the people. These diviners and seers also tracked events, both fortuitous and disastrous. Although they lacked the ability to fully comprehend the celestial mechanics of the planetary system, they were able to visually spot the planets Mercury, Venus, Mars, Jupiter, and Saturn in the heavens. They correlated changes in human emotion and changes in societal events to these planets. They assigned to these planets the names of the various deities revered by the people. They identified and named various star constellations in the heavens and divided the heavens into twelve signs. This was the birth of astrology as we know it today.

Stories of traders benefiting from astrology are also not new. In the early 1900s, esoteric thinkers such as the famous Wall Street trader W.D. Gann reportedly made massive gains when he realized that cycles of astrology bore a striking correlation to financial market action. Gann is most famous for identifying the Saturn/Jupiter cycle which he labelled the Gann Master Cycle. He followed cyclical activity of Jupiter and Neptune when trading wheat and corn futures. He also delved deep into esoteric math, notably square root math which led him to his Square of Nine. The concept of price squaring with time is also a Gann construct. Today many traders and investors attempt to emulate Gann but they do so in a linear fashion, looking for repetitive cycles on the calendar. What they are missing is the astrology component, which is anything but linear.

In the 1930s, Louise McWhirter contributed significantly to financial astrology. She identified an 18.6-year cyclical correlation between the general state of the American economy and the position of the North Node of the Moon in the zodiac. Her methodology extended to include the transiting Moon passing by key points of the 1792 natal birth horoscope of the New York Stock Exchange. She also identified a correlation between price movement of a stock and those times when transiting Sun, Mars, Jupiter and Saturn made hard aspects to the natal Sun position in the stock's natal birth (first trade) horoscope.

The late 1940s saw further advancements in the field of financial astrology when astrologer Garth Allen (a.k.a. Donald Bradley) produced his Siderograph Model based on aspects between the various transiting planets. Each aspect as it occurs is given a sinusoidal weighting as the orb (separation) between the planets varies. Bradley's model is as powerful today as it was in the late 1940s.

As the 1950s dawned, academics at institutions like Yale and Harvard came to dominate the discussion and the cycles of astrology were soon swept aside out of public view. Cyclical analysis was replaced by academic creations like Modern Portfolio Theory and the Efficient Market Hypothesis. These persisted for several decades until coming under severe scrutiny with the 2000 tech bubble meltdown and again with the 2009 sub-prime mortgage crisis which nearly derailed the global economy.

In the past decade, astrology has made a re-appearance. The software designers at Market Analyst/Optuma now have an impressive financial astrology platform built into their charting program. More recently, author and trader Fabio Oreste has published a book on Quantum Lines, a powerful tool based on the work of Einstein, Niels Bohr and Bernhard Reimann.

Think back to the dark days of late 2008 when there was genuine concern over the very survival of the financial market system. This timeframe was the end of an 18.6-year cycle of the North Node traveling around the zodiac. To those players at high levels in the financial system who understood astrology, this period was a prime opportunity to feast off the fear of the investing public and the anxiety of government officials who were standing at the ready with lucrative bailout packages. The market low in March 2009 came at a confluence of a Mars and a Neptune quantum point. Curiously enough, Mars and

Neptune are deemed to be the planetary rulers of the New York Stock Exchange.

Think back to August 2015 and the market selloff that the financial media did not see coming. The reality is this selloff started at a confluence of two events: Venus retrograde event and the appearance of Venus as a morning star after having been only visible as an evening star for the previous 263 days.

Remember the early days of 2016 when Mercury was retrograde and the markets hit a rough patch? Remember the weakness of June 2016 when Venus emerged from conjunction to become visible as an Evening Star?

Do you recall the dire predictions for financial market calamity following the election of Donald Trump to the White House? When the markets instead powered higher, analysts were flummoxed. Venus was making its declination minima right at the time of the US election. Venus declination minima events bear a striking correlation to changes of trend on US equity markets.

What about the early days of 2018 when fear once again gripped the system? Venus was at its declination minimum. Markets reached a turning point in the first week of October 2018 when Venus was agan at a declination low. Add the fact that Venus turned retrograde at the same time and the fear starts to make sense. Markets sold off sharply into mid-December before starting to recover. Sun was conjunct Saturn at this time which correlates strongly to trend changes on equity markets. The North Node had also just changed zodiac signs, an event which also aligns with trend changes.

Markets hit a sudden rough patch in early August 2019. Mercury had just finished a retrograde event and the Federal Reserve cut interest rates as Moon transited a key point on the NYSE 1792 natal horoscope.

US equity markets peaked in late February 2020 and went into total spasm in March 2020. Mercury was retrograde and Mars had just made its declination minimum.

As I finish the edits to this manuscript, the Electoral College votes to determine who will be President of the United States remain undecided. The day of the Election, Mercury finished a retrograde event and heliocentric Jupiter and Saturn were exactly at 0 degrees of separation. To have these two events occur right at the Election date is a rarity.

And so it goes. The cycles continue to unfold as time marches on. The vast majority of people remain unable to comprehend these cycles. But, these cycles are hidden in plain view. All we have to do is look for them. Those who are able to see these cycles can take steps to protect themselves and profit accordingly. Those who remain unaware will continue to ride the emotional roller coaster as their financial planners tell them that investing is for the long term and not to worry.

I personally began to embrace financial astrology in 2012 which was a monumental shift given that my educational background comprises an Engineering degree and an MBA degree. As I pen the text for this 2021 Almanac, I have just completed the requirements for my M.Sc. degree. My approach to astrology is thus heavily slanted towards mathematics and less so towards the classical approach. As you will discover as you turn these pages, I aim to bring as much analytical science into my writings as possible.

This Almanac begins by offering the reader a look at the basic science of astrology. What then follows is an examination of the various cycles that I believe drive the performance of the financial markets. For each type of cycle, I delineate times in the calendar months of 2021 when investors ought to be alert to possible trend changes on the US equity markets. In this age of global connectedness, moves on the S&P 500 are often quickly reflected in other global indices. Chapter 11 then provides a look at various commodity futures and the astro phenomena that influence their price action. Chapter 13 provides a look at the concept of Price Square Time as well as the concept of Quantum Price Lines, a powerful esoteric technique that can be used when applying astrology to making trading and investing decisions.

When applying astrology to trading and investing, it is vital at all times to be aware of the price trend. In my personal experience, the chart indicators developed by J. Welles Wilder are very effective at identifying trend changes. In particular, the DMI and the Volatility Stop are two indicators that should be taken seriously. Lately, I have been using a Japanese construct called the Ichimoku Cloud to identify trend changes. As a trader and investor, look for a change of trend that aligns to an astrology event. When you see the trend change, you should take action. Whether that action is implementing a long position, a short position, an Options strategy, or just tightening up on a stop loss will depend on your personal appetite for risk and on your investment and trading objectives. Using astrology for financial investing is not about taking action at each and every astro event that comes along because not all astro events are powerful enough to induce a change of trend.

This Almanac, which is my seventh such annual publication, is designed to be a resource to help you stay abreast of the various astro events that 2021 holds in store. I am also the author of several other astrology books. In addition, I publish a bi-weekly subscription-based newsletter called *The Astrology Letter*. Through all of my written efforts, I hope to encourage people to embrace financial astrology as a valuable tool to aid in trading and investing decision making.

CHAPTER ONE

Fundamentals

The Sun is at the center of our solar system. The Earth, Moon, planets and various other asteroid bodies complete our planetary system. In addition to the Sun and Moon, there are eight celestial bodies important to the application of astrology to the financial markets. These planets are Mercury, Venus, Mars, Jupiter, Saturn, Uranus, Neptune, and Pluto. Figure 1 illustrates these various bodies and the spatial relation to the Sun on the ecliptic plane. Mercury is the closest to the Sun while Pluto is the farthest away.

The Ecliptic and the Zodiac

The various planets and other asteroid bodies rotate 360 degrees around the Sun following a path called the ecliptic plane. As shown in Figure 2, Earth (and its Equator) is slightly tilted (approximately 23.45 degrees) relative to the ecliptic plane. Projecting the Earth's equator into space produces the celestial equator plane. There are two points of intersection between the ecliptic plane and celestial equator plane. Mathematically, this makes sense as two non-parallel planes must intersect at two points. These points are commonly called the vernal equinox (occurring at March 20th) and the autumnal equinox (occurring at September 20th). You will recognize these dates as the first day of Spring and the first day of Fall, respectively. Dividing the ecliptic plane into twelve equal sections of 30 degrees results in what astrologers call

7

the zodiac. The twelve portions of the zodiac have names including Aries, Cancer, Leo and so on. Ancient civilizations looking skyward identified patterns of stars called constellations that align to these twelve zodiac divisions. If these names sound familiar, they should. You routinely see all twelve names in the daily horoscope section of your morning newspaper.

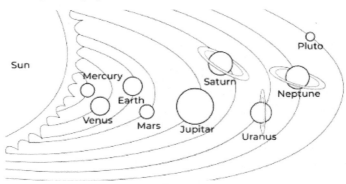

Figure 1
The Planets
(image: https://harikerja.com/)

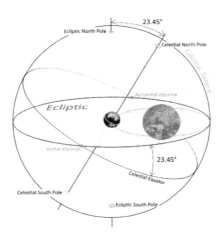

Figure 2
The Ecliptic
(image: www.whenthestarscomeout.com)

The Glyphs

Figure 3 illustrates the symbols that appear in the twelve segments of a zodiac wheel. The segments are more properly called signs. These symbols are called glyphs.

The starting point or zero degree point of the zodiac wheel is the sign Aries, located at the vernal equinox of each year. The vernal equinox is when, from our vantage point on Earth, the Sun appears at zero degrees Aries. The autumnal equinox is when, from our vantage point on Earth, the Sun appears at 180 degrees from zero Aries (0 degrees of Libra).

Figure 3
The Zodiac Wheel
(image: https://www.elsaelsa.com/astrology/zodiac-sign-glyphs/)

The various planets are also denoted by glyphs, as shown in Figure 4.

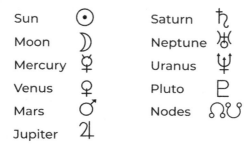

Figure 4
The Glyphs
(image: http://earther-rise.com/interpretation-guide/planet-glyphs)

Declination

As the various celestial bodies make their respective journeys around the Sun, they can be seen to move above and below the celestial equator plane. This movement is termed declination. Celestial bodies experience declinations of up to about 25 degrees above and below the celestial equator plane.

Mercury, Venus, and Mars endure frequent changes in declination due to the gravitational force of the Sun. Planets like Jupiter, Saturn, Neptune, Uranus and Pluto also experience declination changes but these changes are slower to evolve. As this Almanac will illustrate, changes in the declination of a celestial body (most notably Mars and Venus) can affect the financial markets.

Parallel and Contra-Parallel

Declination can be viewed one planet at a time or by pairs of planets. Let's suppose that at a particular time Mars can be seen as being 10 degrees of declination above the celestial equator and at that same time Venus is at 9 degrees of declination. Let's further suppose that we allow for up to 1.5 degrees tolerance in our measurement of declinations. We would say these two planets were at parallel declination. Let's take another example and suppose that at a given time Jupiter was at 5 degrees of declination above the celestial equator and at that same time period Pluto was at 6 degrees declination below the celestial equator. Again, let's allow for up to 1.5 degrees of tolerance. We would say that Jupiter and Pluto were at contra-parallel declination. As this Almanac will show, parallel and contra-parallel events can influence the financial markets.

The Moon

Just as the planets orbit 360 degrees around the Sun, the Moon orbits 360 degrees around the Earth. The Moon orbits the Earth in a plane of motion called the lunar orbit plane. This plane is inclined at about 5 degrees to the ecliptic plane as Figure 5 shows. The Moon orbits Earth with a slightly elliptical pattern in approximately 27.3 days, relative to an observer located on a fixed frame of reference such as the Sun. This time period is known as a sidereal month. However, during one sidereal month, an observer located on Earth (a moving frame of reference) will revolve part way around the Sun. To that Earth-bound observer, a complete orbit of the Moon around the Earth will appear longer than the sidereal month at approximately 29.5 days. This 29.5 day period of time is known as a synodic month or more commonly a lunar month. The lunar month plays a key role in applying astrology to the financial markets as will be detailed throughout this book.

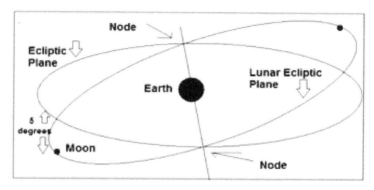

Figure 5
(image: http://www.physics.unlv.edu/~jeffery/astro/ial/ial_003.html)

The Nodes

A mathematical construct related to the Moon and central to financial astrology is the Nodes. The Nodes are the points of intersection between the Earth's ecliptic plane and the Moon's ecliptic plane. Figure 5 also illustrates the Nodes. In astrology, typically only the North Node is referred to. (The North Node forms the basis for the McWhirter Method which will be discussed later in this book).

Ascendant, Descendant, MC and IC

As the Earth rotates on its axis once in every 24 hours, an observer situated on Earth will detect an apparent motion of the constellation stars that define the zodiac. To better define this motion, astrologers apply four cardinal points to the zodiac, almost like the north, south, east and west points on a compass. These cardinal points divide the zodiac into four quadrants. The east point is termed the Ascendant and is often abbreviated Asc. The west point is termed the Descendant and is often abbreviated Dsc. The south point is termed the Mid-Heaven (from the Latin Medium Coeli) and is often abbreviated MC or MH. The north point is termed the Imum Coeli (Latin for bottom of the sky) and is abbreviated IC. Figure 6 illustrates the placement of these cardinal points on a typical zodiac wheel. The importance of the Ascendant and Mid-Heaven will be emphasized in more detail when the McWhirter Method is discussed.

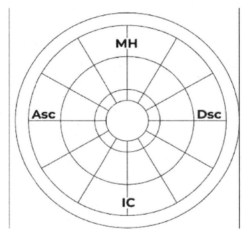

Figure 6
Cardinal Points

Geocentric and Heliocentric Astrology

Astrology comes in two distinct varieties – geocentric and heliocentric.

In geocentric astrology, the Earth is the vantage point for observing the planets as they pass through the signs of the zodiac. Owing to the different times for the planets to each orbit the Sun, an astrologer situated on Earth would see the planets making distinct angles (called aspects) with one another and also with the Sun. The aspects that are commonly used in astrology are 0, 30, 45, 60, 90, 120, 150 and 180 degrees. In financial astrology, it is common to refer to only the 0, 90, 120 and 180 degree aspects.

In heliocentric astrology, the Sun is the vantage point for observing the planets as they pass through the signs of the zodiac. An observer positioned on the Sun would also see the orbiting planets making aspects with one another.

To identify these aspects, astrologers use *Ephemeris Tables*. For geocentric astrology, the *New American Ephemeris for the 21st Century* is commonly used. For heliocentric astrology, the *American Heliocentric Ephemeris* is a good resource.

For faster aspect determination, two excellent software programs available are Millenium Trax produced by AIR Software and Solar Fire Gold produced by software company Astrolabe. My preference is the Solar Fire Gold product. I also use a market platform called OPTUMA/Market Analyst. This brilliant piece of software, originally developed in Australia, allows the user to generate end of day price charts for equities and commodities from a multitude of exchanges and then overlay various astrological aspects and occurrences onto the chart. As your journey into astrology deepens, you might be tempted to spend the money to acquire this software program.

Synodic and Sidereal

The vantage point of either Earth or Sun leads to two more concepts, synodic and sidereal. These descriptors were discussed earlier in the context of the Moon. To an earth-bound observer, a sidereal time period is the time between two successive occurrences. That is, how many days does it take for Sun passing Pluto on the zodiac wheel to again pass Pluto? To a Sun-bound observer, a synodic time period is the number of days (or years) it takes for a planet to orbit the Sun. These time frames play roles in assessing market cycles as will be discussed in this Almanac. The data in Figure 7 presents synodic and sidereal data.

Planet	Synodic Period	Sidereal Period
Mercury	116 days	88 days
Venus	584 days	225 days
Mars	780 days	1.9 years
Jupiter	399 days	11.9 years
Saturn	378 days	29.5 years
Uranus	370 days	84 years
Neptune	368 days	164.8 years
Pluto	367 days	248.5 years

Figure 7
Synodic and Sidereal Data

Retrograde

Think of the planets orbiting the Sun as a group of cars travelling around a racetrack. Consider what happens as a fast moving car approaches a slower moving car from behind. At first, all appears normal. An observer in the fast moving car sees the slower moving car heading in the same direction. Gradually, the observer in the fast car sees that he will soon overtake the slow car. For a brief moment in time as the fast car overtakes the slower car the observer in the fast car notices that the slower car appears to stand still and even move backwards. Of course the slow car is not really standing still. This is simply an optical illusion.

These brief illusory periods are what astrologers call retrograde events. To ancient societies, retrograde events were of great significance as human emotion was often seen to be changeable at these events. Is it possible that our DNA is hard-wired such that we feel uncomfortable at retrograde events?

From the vantage point of an observer on Earth, there will be three or four times during a year when Earth and Mercury pass by each other on this celestial racetrack. There will be one time every couple of years when Earth and Venus pass each other. There will be one time every two years when Earth and Mars pass each other. Retrograde events all too often will see a short term trend change develop.

Elongation and Conjunction

From an observer's vantage point on Earth, there will also be times when planets are seen to be at maximum angles of separation from the Sun. These events are what astronomers refer to as maximum easterly and maximum westerly elongations. These events definitely have a correlation to trend changes on markets.

Mercury and Venus are closer to the Sun than is the Earth. From our vantage point on Earth, there will be times when Mercury and Venus are between the Earth and the Sun. Likewise, there will be times when the Sun is between the Earth and Mercury or Venus. On the zodiac wheel, the times when Mercury or Venus are in the same zodiac sign and degree as the Earth are what astronomers call conjunctions.

An Inferior Conjunction occurs when Mercury or Venus is between Earth and the Sun.

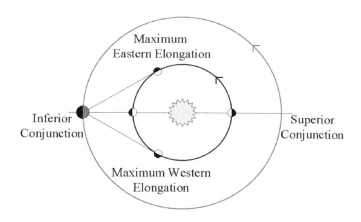

Figure 8
Superior and Inferior Conjunction
(image: http://www.astronomy.ohio-state.edu/~pogge/Ast161/Unit2)

A Superior Conjunction occurs when the Sun is between Earth and Mercury or Venus. Figure 8 illustrates the concept of elongation and conjunction.

Conjunction events occur on either side of retrograde events. For example, in 2018 Venus was retrograde from October 5 to November 15. Its actual Inferior Conjunction was recorded on October 26.

After Venus has been at Inferior Conjunction, it will be visible as a Morning Star. After it has been at Superior Conjunction, it will be visible as an Evening Star. Venus was at Superior Conjunction on March 28, 2013 (8 Aries), October 25, 2014 (1 Scorpio), June 6, 2016 (16 Gemini), and January 8, 2018 (18 Capricorn). Venus was at Inferior Conjunction on June 6, 2012 (15 Gemini), January 11, 2014 (21 Capricorn), August 15, 2015 (22 Leo), March 25, 2017 (4 Aries), and October 26, 2018 (3 Scorpio). Plot these groups of Superior Conjunction events on a zodiac wheel. Note how they can be joined to form a 5-pointed star called a pentagram. Likewise, the Inferior Conjunction events can be plotted and joined to form a pentagram. Such is the elegance and mystique of the cosmos.

CHAPTER TWO

The Master Cycle

W.D. Gann closely followed the cycles of Jupiter and Saturn. To an observer situated on the fixed vantage point of the Sun, Jupiter would be seen orbiting the Sun in about 12 years and Saturn in just over 29 years. Gann interpreted these orbital cycles one step further and noted that every 20 or so years heliocentric Jupiter and Saturn were at conjunction, in a particular sign of the zodiac, separated by zero degrees. This is what he called the Master Cycle.

Gann further noted that the financial markets were affected by this Master Cycle. The market crash of 1901 aligned to a conjunction of these two outer planets. In 1920, the U.S. economy encountered a recession, and in 1921 the financial markets reached a low point again at the conjunction of these two heavy-weight planets. In 1941, the markets recorded a low point a couple months after another Jupiter-Saturn conjunction. The early-1960s should have seen another major turning point at the end of a Master Cycle. There was a drawdown in the U.S. equity markets that historians now call the Kennedy Slide, but it did not align perfectly to the Master Cycle. As it turns out, this one exception to the Master Cycle was affected by the far more powerful Cardinal Cross.

Picture a rectangle with its parallel sides and parallel ends. Now, place this rectangle inside of the zodiac. The corner points of the rectangle are pairs of planets. This 1965 Cardinal Cross involved eight celestial bodies (Neptune was the odd planet out). With the Dow Jones

17

Average at near 8000 points, the Cardinal Cross marked a significant turning point for the US equity market. It would not be until 1995 when the US equity market would again test the 8000 level.

In the Spring of 1981, Jupiter and Saturn recorded a Master Cycle conjunct event. A handful of months later, the U.S. equity markets recorded a very important low that marked the onset of a massive bull market that ran until the next conjunction event in mid-2000. If you were active in the markets then, you may have painful memories of the tech-bubble and of money lost in your brokerage account. Figure 9 illustrates how the S&P 500 was making its peak as Jupiter and Saturn were making their conjunction. What happened after this conjunction is even more interesting. The market ground painfully lower until late 2002. Rallies along the way were systematically crushed.

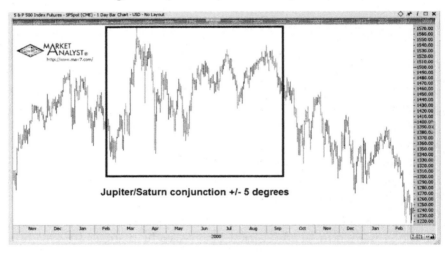

Figure 9
Jupiter / Saturn Conjunction in 2000

As I finish this manuscript in November 2020, Saturn and Jupiter have just made their zero degree conjunction. For many traders and investors, 2020 has potential to leave lasting memories. This conjunction event could be a powerful one as the heliocentric zodiac chart shows 7 celestial features crammed into a 90 degree segment of the zodiac. That is a significant amount of planetary energy. Will the aftermath of this conjunction see the market grind lower for an extended time as it did following the 2000 conjunction? Or, will this conjunction see the equity markets disconnect from reality and surge even higher driven by even more aggressive quantitative easing and fiat money stimulus?

By the time this book comes off the press, the Saturn-Jupiter conjunction will have completed itself and the markets will have signaled which direction they intend to go in. The end of one Master Cycle and the start of another one always heralds opportunities for the alert investor. At the end of the last Master Cycle, Facebook and Twitter had not yet been founded. Cloud computing was an expression that did not exist. Going to a retail shop on the street corner to legally buy cannabis was not even dreamed of. Renewable energy was in its infancy. Global wind power generation in 2000 was 17,000 MW. Today the number stands at 650,000 MW.

I am curious to see what market sectors become favorable as this new cycle begins. Will Gold finally gallop higher as precious metals pundits have been calling for? What sectors will fall out of favor? Will the conventional oil industry wither and die? What renewable energy technologies will emerge in its place? Will the US Dollar cease to be the global reserve currency? Will China continue to dominate global manufacturing or will the further rise of China be kept in check with tariffs and military posturing? Will the economy emerge intact from the COVID pandemic or will the early 2021 event of Mars passing the Mars and Neptune locations in the 1776 USA natal chart trigger stock market panics? Will any weakness be exacerbated by Neptune being opposite the 1776 natal Neptune with Node being opposite the 1776 natal Gemini? Will society draw closer together or will people turn on each other. Will Uranus passing through the sign of Gemini from 2025 to 2033 put America at war on foreign soil or at civil war at home?

CHAPTER THREE
The 18.6 Year Cycle

There exists another longer cycle that more people ought to be aware of. This cycle was first written about in the 1930s by a mysterious figure called Louise McWhirter. I say mysterious because in all my research I have neither come across any other writings by her nor have I found reference to her in other manuscripts. I am almost of the opinion that the name was a pseudonym for someone seeking to disseminate astrological ideas while remaining anonymous.

The Moon orbits the Earth in a plane of motion called the lunar ecliptic. Two planes that are not parallel will always intersect at two points. The two points where the lunar ecliptic intersects the plane of motion of planet Earth are termed the North Node and South Node. McWhirter recognized that the transit of the North Node of the Moon around the zodiac wheel takes 18.6 years and that the Node progresses in a backwards motion through the zodiac signs. Through examination of copious amounts of economic data provided by Leonard P. Ayers of the Cleveland Trust Company, McWhirter was able to conclude that when the North Node moves through certain zodiac signs, the economic business cycle reaches a low point and when the Node is in certain other signs, the business cycle is at its strongest.

This line of thinking is still with us today. The most notable authority embracing this cycle is Australian economist Fred Harrison. In his published works, he discusses this long economic cycle going back to the Industrial Revolution. But, to maintain respect in academia, he stops just shy of stating a connection to astrology.

McWhirter was able to discern the following from her research:

- As the Node enters Aquarius, the low point of economic activity is reached

- As the Node leaves Aquarius and begins to transit through Capricorn and Sagittarius, the economy starts to return to normal

- As the Node passes through Scorpio and Libra, the economy is functioning above normal

- As the Node transits through Leo, the high point in economic activity is reached

- As the Node transits through Cancer and Gemini, the economy is easing back towards normal

- As the Node enters the sign of Taurus, the economy begins to slow

- As the Node enters Aquarius, the low point of economic activity is reached and a full 18.6 year cycle is completed.

McWhirter further observed some secondary factors that could influence the tenor of economic activity in a good way, no matter which sign the Node was in at the time:

- Jupiter being 0 degrees conjunct to the Node

- Jupiter being in Gemini or Cancer

- Pluto being at a favorable aspect to the Node

McWhirter also observed some secondary factors that can influence the tenor of economic activity in a bad way, no matter which sign the Node was in at the time:

- Saturn being 0, 90 or 180 degrees to the Node

- Saturn in Gemini or Cancer

- Uranus in Gemini

- Uranus being 0, 90 or 180 degrees to the Node

- Pluto being at an unfavorable aspect to the Node.

In the Summer of 2019, there were media rumblings about an imminent recession. But, several large investment banking firms in New York weighed in with a collective dissenting opinion that pushed any possibility of recession off into the future. I perceive this to be evidence that the large money firms in New York are paying close attention to the 18.6-year cycle. Soon enough, the obedient media talking heads were echoing this non-recessionary sentiment.

So, what has 2020 delivered? North Node was in the sign of Cancer in early 2020. As I pen this manuscript, the Node is working its way through Gemini. McWhirter's work suggests movement of the Node through Cancer and Gemini will align to a slowing of the economy back toward normal levels of activity. If each of us looks around in our communities, I think we can all agree that economic activity has eased, especially in response to COVID 19.

As the Node moves through Taurus and Aries, the prospect of recession will loom ever larger. Recession will manifest in a big way starting in late 2024. By 2026, the Node will be in Aquarius to mark the end of the 18.6 year cycle and very likely the depths of another financial crisis. Uranus will be in Gemini and 90 degrees to the Node. This astro positioning warns of a very negative time. One other astro cycle that I follow is even indicating the U.S.A. being at war at the time the 18.6 year cycle ends. The 1776 natal horoscope for the USA has Uranus in the sign of Gemini. In 2026, Uranus will again enter the sign of Gemini and by mid-2027 will be exactly conjunct to the 1776 Uranus natal position at 8 of Gemini. But these events are still a good four years distant, so try not to worry too much. Just be aware. Plan accordingly. Take steps to protect yourself and your family. I will leave it to you to decide what 'taking steps' means.

CHAPTER FOUR

Venus Cycles

Cycles of Venus play a key role in financial astrology. Venus orbits the Sun in 225 days relative to an observer standing at a fixed venue like the Sun. To an observer situated on Earth (a moving frame of reference), Venus appears to take 584 days to orbit the Sun. During this 584 days, that same observer on Earth will note periods of time when Venus is not visible in the early morning or evening sky. This is because the planet is between Earth and Sun in its orbital journey. This is the Inferior Conjunction. As Venus slowly moves out of this conjunction, it will become visible as the Morning Star. During that part of its journey when Venus is 180 degrees opposite Earth, it is said to be at Superior Conjunction. As it moves out of this conjunction, it becomes visible as the Evening Star.

Venus was at Superior Conjunction on March 28, 2013 (8 Aries), October 25, 2014 (1 Scorpio), June 6, 2016 (16 Gemini) and January 8, 2018 (18 Capricorn). Venus was at Inferior Conjunction on June 6, 2012 (15 Gemini), January 11, 2014 (21 Capricorn), August 15, 2015 (22 Leo), March 25, 2017 (4 Aries), and October 26, 2018 (3 Scorpio). If you plot these groups of Superior Conjunction events on a zodiac wheel you will note how they can be joined to form a 5-pointed star called a pentagram. Likewise, the Inferior Conjunction events can be plotted and joined to form a pentagram. Such are the mysteries of our cosmos.

As Venus orbits around the Sun following the ecliptic plane, it moves above and below the plane. The high points and low points made during this travel are termed declination maxima and minima.

In early 2018, a Venus Superior Conjunction and declination minimum was followed closely by a steep 300 point sell-off on the S&P 500. A significant trend change immediately preceded an Inferior Conjunction and declination minimum in October 2018. This was followed by an acceleration of trend to the downside.

In July 2019, talk of a recession and an impasse with China over trade disputes caused a sharp sell-off on the S&P 500. But this sell-off event was nothing more than a Venus Superior Conjunction combined with a Venus declination maximum.

A Venus Superior Conjunction in early December 2019 failed to deliver a meaningful reaction. The Federal Reserve was busy ramping up liquidity to the banking system which had the effect of dampening the astrology cycles.

In early May 2020, Venus made a declination maximum and the S&P 500 exhibited a short, sharp 200 point drop.

The Inferior Conjunction event on June 3 delivered a 200 point drop to the S&P 500. But again the Federal Reserve came to the rescue with more fiat liquidity.

- The next Superior Conjunction of Venus will be March 26 of 2021

- Venus will be at a declination low point in early January, 2021

- Venus will be at a declination maximum in early June, 2021

- Venus will be at a declination minimum in early November 2021.

As each of these Venus events approach, it is highly advisable to be alert for sudden market moves, higher or lower, which could impact your investment portfolio. Watch the S&P 500 relative to the 21 and 34 day moving average for evidence of a trend change. Also, watch the S&P 500 price chart for evidence of a trend indicator (such as the Wilder Volatility Stop) giving evidence of a trend change.

Another cyclical event pertaining to Venus is its retrograde events. When discussing the basics of astrology in a previous chapter, I used the analogy of cars on a racetrack passing each other to explain retrograde.

To further help understand the science of Venus retrograde, consider the diagram in Figure 10.

In 30 days of time, planet Earth (shown as the larger circles in the diagram) will travel 30 degrees of the zodiac (from point 1 to point 2).

But, Venus is a faster mover. In the same 30 days of time, Venus (shown as the smaller circles) will travel through about 42 degrees of the zodiac, passing by Earth in the process. From our vantage point here on Earth, as Venus is setting up to pass Earth, we see Venus in the sign of Sagittarius. As Venus completes its trip past Earth, we see it in the sign of Scorpio.

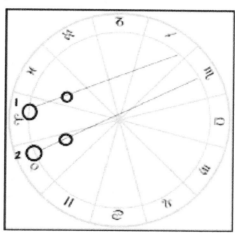

Figure 10
The concept of Venus retrograde

Venus appears to have moved backwards in zodiac sign as it has passed by Earth. This is the concept of retrograde. To the ancients who did not fully understand how the cosmos worked, it must have been awe-inspiring to see a planet move backwards in the heavens relative to their constellation stars.

There is a curiously strong correlation between equity markets and Venus retrograde. Sometimes Venus retrograde events encompass a sharp market inflection point. Sometimes a market peak or bottom will follow closely behind a retrograde event. Sometimes a peak or bottom will immediately precede a retrograde event. When you know a Venus retrograde event is approaching, use a suitable chart technical indicator such as DMI or Wilder Volatility Stop to determine if the price trend is changing.

Figure 11 illustrates price behavior of the S&P 500 Index during May 2020 as Venus was retrograde.

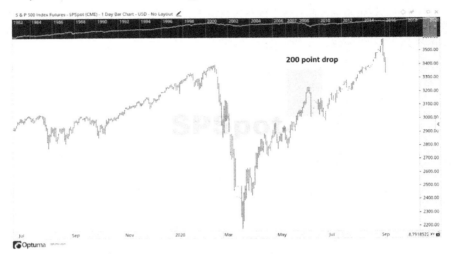

Figure 11
S&P 500 Index and Venus retrograde

Knowing that the potential exists for sizeable moves, aggressive traders can avail themselves of these retrograde correlations. Less aggressive investors may simply wish to place a stop loss order under their positions to guard against sharp price pullbacks.

As a further example, Figure 12 illustrates price performance of the S&P/ASX Australian 200 Index. The May 2020 Venus retrograde event delivered a short, sharp blow to the index. At the time of writing, the highs made during retrograde event have yet to be challenged.

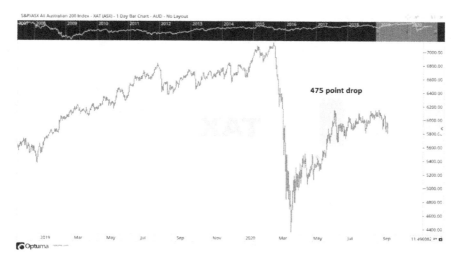

Figure 12
Venus retrograde and the ASX 200 Index

For 2021, Venus will be retrograde from mid-December through to January 2022.

CHAPTER FIVE

Mercury Cycles

Mercury is the smallest planet in our solar system. It is also the closest planet to the Sun. As a result of its proximity to the powerful gravitational pull of the Sun, Mercury moves very quickly, completing one sidereal cycle of the Sun in 88 days.

Scientists at NASA conclude that Mercury has a di-polar magnetic field. This field is strong enough to deflect solar wind particles that have emanated from the Sun. These deflected solar winds then carry on towards the Earth where they presumably affect the magnetic field in our cosmos. Whether this affected magnetic field alters human emotion (and thereby the markets) remains uncertain.

Scientists have also determined that Mercury has an eccentric orbit in which its distance from the Sun will range from 46 million kms to 70 million kms. When Mercury is nearer to the Sun (ie 46 million kms away), it is moving at its fastest (~56.6 kms per second). When Mercury is farther from the Sun (ie 70 million kms away), it is moving slower (~38.7 kms per second).

Related to Mercury's orbit is its elongation. As discussed earlier, elongation refers to the angle between a planet and the Sun, using Earth as a reference point. Figure 13 illustrates the notion of elongation.

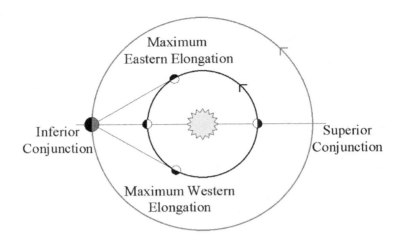

Figure 13
Elongation of Mercury

(image: http://www.astronomy.ohio-state.edu/~pogge/Ast161/Unit2)

In 2020, Mercury was at its greatest easterly elongation February 10, June 3 and October 1. Greatest westerly elongation occurred March 24, July 22, and November 10.

Figure 14 illustrates some of these 2020 events overlaid on a chart of the S&P 500. Note how a greatest easterly elongation event and a greatest westerly elongation event defined the high and low of the COVID-related market sell-off. Another greatest easterly elongation event helped to define (along with Venus retrograde) the early June short spell of weakness. Is this all just scientific coincidence? Or, is this a stunning example of powerful players using astrology to move the markets when they desire? It appears to me that the latter situation is the one at work. All we can do as traders and investors is to mark the elongation dates on our calendars and pay close attention to the market trend.

For 2021, Mercury will be at its greatest easterly elongation January 23, May 17, and September 13. Greatest westerly elongation events will occur March 6, July 4, and October 25.

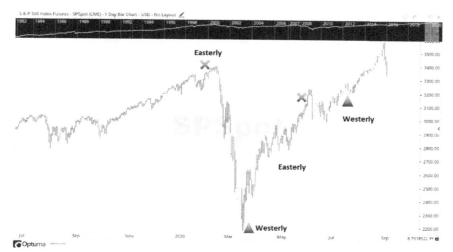

Figure 14
2020 Mercury Elongation events

In addition to times of maximum elongation, there will be retrograde events. Mercury retrograde is probably one of the most potent planetary influences for investors and traders to be aware of. We often hear about Mercury retrograde events in mundane astrology. Classical astrologers will tell clients to not sign important contracts during Mercury retrograde, to not cross the street, to not leave their houses and so on. While I tend to ignore this mundane talk, I have noticed a striking correlation between financial market behavior and Mercury retrograde events.

To understand the science of Mercury retrograde, consider the diagram in Figure 15.

In 30 days of time, planet Earth (shown as the larger circles in the diagram) will travel 30 degrees of the zodiac (from point 1 to point 2). But, Mercury is a faster mover. In the same 30 days of time, Mercury (shown as the smaller circles) will travel through about 120 degrees of the zodiac (point 1 to point 2), passing by Earth in the process. From our vantage point on Earth, as Mercury is setting up to pass Earth, we see Mercury in the sign of Aries. These sign determinations are made by extending a line from planet Earth through Mercury to the outer edge of the zodiac wheel. As Mercury completes its trip past Earth, we see it in the sign of Capricorn. In other words, the way we see it here on Earth, Mercury has moved backwards in zodiac sign as it passed Earth.

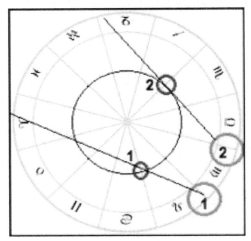

Figure 15
The concrpt of Mercury retrograde

Sometimes Mercury retrograde events encompass a sharp market inflection point. Sometimes a market peak or bottom will follow closely behind a retrograde event; sometimes a peak or bottom will immediately precede a retrograde event. When you know a Mercury retrograde event is approaching, use a suitable chart technical indicator such as DMI or Wilder Volatility Stop to determine if the price trend is changing.

Figure 16 illustrates 2019 and 2020 Mercury retrograde events overlaid on a chart of the S&P 500.

Figure 16
S&P 500 and Mercury retrograde

Mercury retrograde events can often be also seen influencing commodity markets. Figure 17 illustrates the correlation between Copper futures prices and Mercury retrograde. While the average investor may not be aggressively trading Copper futures, this correlation could be used to better manage price risk of copper-related mining stocks in an investment portfolio. Looking at the following chart of Copper futures, one can see the price high in March 2019 aligns to a Mercury retrograde event. A reversal of trend again occurred in July 2019 at a Mercury retrograde event. More recently in July 2020, Mercury retrograde halted a sharp rally.

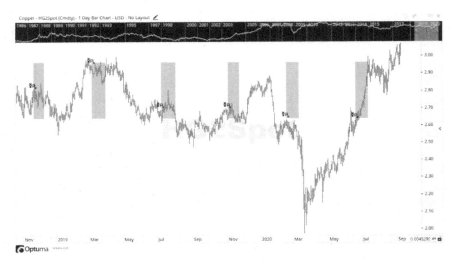

Figure 17
Mercury retrograde and Copper futures

Figure 18 illustrates Gold futures prices overlaid with Mercury retrograde events in 2019 and 2020. In past years when the price trend on Gold varied from positive to sideways to negative, it was easy to spot the influence of Mercury retrograde. I am not seeing such a strong correlation in more recent price action because the trend on Gold price has been largely to the upside.

Figure 18
Mercury retrograde and Gold futures

Whatever country you happen to reside in, check the major stock market index for that country to determine if there is a correlation to Mercury retrograde. If there is, be alert to trend changes at the retrograde events. If trading commodity futures such as Gold or Copper, Mercury retrograde will be most apparent if the prevailing price trend is something less than strongly positive.

For 2021, Mercury will be:

- retrograde from January 30 through February 19

- retrograde from May 31 through June 21

- retrograde from September 28 through October 16.

CHAPTER SIX
Professor Weston's Cycles

At the beginning of this book, I suggested that the markets were a series of cycles all strung together over time. Thanks to a fortuitous discovery made late one night in 2017, I now understand how these cycles can be joined. On that fateful night, whilst scouring the Internet looking for old astrology manuscripts for sale, I came upon a white paper written in 1921 by a mysterious person from Washington, D.C. going by the name of Professor Weston. I paid $50 for this document, and I am thrilled that I did. Never in my research travels to various libraries have I come across this name. I cannot find any other publications by him, although I understand there might be two more white papers out there somewhere. Who exactly he was, I will likely never know; another one of those figures who emerged to write his ideas down and then vanished into the ether.

In his day, Weston analyzed copious amounts of data from the Dow Jones Average. He applied cosine Fourier mathematics and he came up with a general set of rules for the Dow. I have checked his model against the Dow Jones and I find that his model is still accurate.

Perhaps Weston knew W.D. Gann personally. Perhaps he just knew of him. In any case, Weston followed the 20-year Gann Master Cycle cycle of Jupiter and Saturn.

He further broke this long cycle into two components of 10 years. He believed investors can expect:

- a 20-month market cycle to begin in November of the 1st year of the 10-year cycle

- another 20-month cycle to begin in November of the 5th year of the 10-year cycle

- 28-month cycles to begin in July of the 3rd and 7th years of the 10-year cycle

- a 10-month cycle to begin in November of the 9th year of the 10 year cycle

- a 14-month cycle to begin in September of the 10th year of the 10-year cycle.

To put all this into perspective, the new Master Cycle will begin on October 30, 2020 as heliocentric Jupiter and Saturn make a 0 degree aspect. The entire cycle will then run until November 2040.

Following Weston's methodology:

- the first 20-month cycle will start in November 2020 and go to until July 2022

- a 28-month cycle will run from July 2022 through November 2024

- a 20-month cycle will then follow until July 2026

- a 28-month cycle will run July 2026 through November 2028

- a 10-month cycle will then run through until September 2029

- lastly, a 14-month cycle will last until November 2030.

Weston also postulated that in the various years of a 10-year segment of the overall Master Cycle, there would be market maxima as listed in Figure 19.

Weston calculated these events using cycles of Venus. He argued that the 16th Harmonic of a 10-year period (120 months) was actually the heliocentric time it takes for Venus to orbit the Sun (120 x 30 / 16 = 225 days).

After the new Cycle begins in November 2020, Weston cautions to be alert for market maxima in March 2021 and in October 2021.

Year of Cycle	Maxima	Maxima
1	March	October
2		May
3	January	September
4	April	November
5	May	November
6		June
7	January	September
8		June
9	April	
10	February	August

Figure 19
Weston's Secondary Cycles

What amazes me is Weston's predictions for year 10, namely a maxima in February and another in August. In early 2020 (the 10th year of the last half of a Master Cycle), the equity markets peaked in February as fears over COVID-19 began to manifest. Massive fiat money stimulation then drove the markets from their March lows towards an over-valued maximum in late August. The markets truly are an overlapping array of cycles, just as Weston postulated back in the 1920s. The question that remains is, do these cycles just unfold in and of themselves? Or, do manipulative players working behind the curtain play a role?

To illustrate Weston's cycles, consider the 28-month cycle, which Weston said would run from July 2016 to November 2018 (the month of the Congressional mid-term elections in the USA). Figure 20 shows how the end of this 28-month cycle aligned to a market sell-off. Was it the election outcome that influenced the market or was it the end of a short-term market cycle?

From the end of the 28-month cycle in November 2018, a Weston 10-month cycle was to run to September 2019 and an ensuing 14-month cycle to November 2020 and the start of a new Master Cycle.

Figure 21 illustrates further. The one observation to take away from Figure 21 is the need to allow just a bit of flexibility to Weston's start and end times.

Figure 20
Weston's 28-month cycle

Figure 21
Weston's 10 and 14-Month Cycles

I have taught myself how to replicate Weston's cosine mathematics and I am finding fairly robust predictive abilities when I apply the model to Gold prices and other commodity futures. It is important to take adequate time to properly recognize the individual cycles that drive price action on a particular commodity or index. These individual cycles are what form the basis for the cosine calculations.

CHAPTER SEVEN

Sacred Cycles

Shorter cycles related to religion have a curious way of intersecting with financial market turning points. One religious concept is that of Shmitah which is rooted in the Hebrew Bible. I learned of the Shmitah from the writings of Rabbi Jonathan Cahn. On the surface, Cahn appears to be an average ordinary Rabbi from New Jersey, USA. But behind the scenes, he has done a masterful job of applying Shmitah to the financial markets.

In the book of Exodus (Chapter 23, verses 10-11), it is written: *"You may plant your land for six years and gather its crops. But during the seventh year, you must leave it alone and withdraw from it."*

In the book of Leviticus (Chapter 25, verses 20-22), it is written: *"And if ye shall say: 'What shall we eat the seventh year? Behold, we may not sow, nor gather in our increase'; then I will command My blessing upon you in the sixth year, and it shall bring forth produce for the three years. And ye shall sow the eighth year, and eat of the produce, the old store; until the ninth year, until her produce come in, ye shall eat the old store."*

Breaking these Biblical statements down into simple-to-understand terms means that every 7th year something will happen on the financial markets.

The first Shmitah year in the modern State of Israel was 1951-52. Subsequent Shmita years have been 1958–59, 1965–66, 1972–73, 1979–80, 1986–87, 1993–94, 2000–01, 2007–08, and 2014-15. The next Shmita year will be 2021-2022. A Shemitah Year starts in the month of

Tishrei (the first month of the Jewish civil Calendar) and ends in the month of Elul. The Gregorian calendar equivalent will have the 2021 Shmitah year starting in the September-October timeframe.

The chart in Figure 22 illustrates the S&P 500 with some recent Shmitah years overlaid. The select few who understand Shmitah would have profited handsomely from these moves on the S&P 500.

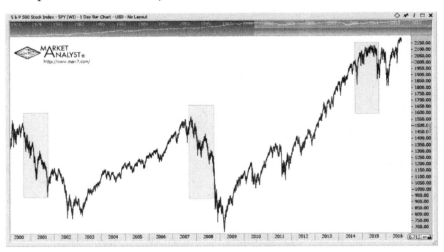

Figure 22
S&P 500 and Shmitah years

Shmitah years are not always about the equity markets. The chart in Figure 23 illustrates Oil prices with Shmitah years overlaid. Those who understood and used Shmitah as an investing strategy made serious money on the Crude Oil market in 2007 and again in 2014.

Figure 23
Crude Oil and Shmitah years

Referrring back to the Exodus and Leviticus Biblical passages, the message is there shall be no crop in the Shmitah year. In the year after the Shmitah, people shall live on the bounty of the crop produced in year 6, the year immediately prior to Shmitah. In the 9th year, the newly planted crop will be abundant.

The suggestion for here and now is, live off the bounty reaped during 2020, for there will not be a bountiful crop in 2021 nor in 2022. We can all see the massive fiat money stimulus that drove the US equity markets to ridiculous levels in 2020 following the initial COVID panic. A bounty indeed for those institutions who availed themselves of the easy Fed QE money.

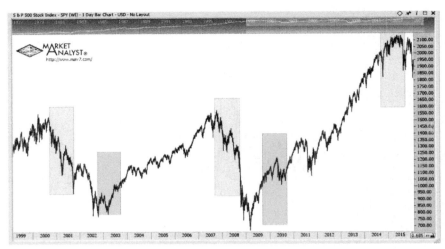

Figure 24
The Aftermath of Shmitah years

In Figure 24, which illustrates price action on the S&P 500, note the gap following the 2007-2008 Shmitah event. This gap is followed by another rectangle to mark what effectively is the second year after Shmitah. The key point is in the year immediately after Shmitah the market can fall and take out the price lows of the Smhitah year. The earlier-cited Biblical passages advise that we are to be living off the produce harvested (financial gains made) prior to Shmitah. Once the post-Shmitah period has elapsed, Figure 24 shows that the market can regian its footing and start to advance again.

The period September 2021 through September 2022 will be a Shmitah year. The post Shmitah year will conclude by September 2023. Structure your trading and investing activity accordingly. Strictly adhering to the Shmitah cycles, profits ideally should have been taken by the September 2020 period. Plan to re-enter the markets in a serious way in September 2023. Along the way, look for only short term trading of oversold opportunities across various segments of the market.

The big question is — what the theme of the 2021 Shmitah will be? Will it be a weakening US Dollar? Will it be surging Gold prices? Will it be increasing interest rates? Will it be a commodity super-cycle?

In addition to the Shmitah cycle, there are certain dates from the Hebrew calendar that Rabbi Cahn says can have a strong propensity to align with swing highs and lows on the New York Stock Exchange. Rabbi Cahn advises to pay close attention to four particular dates from the Hebrew calendar:

- The 1st Day of the month of Tishrei marks the start of the Jewish civil calendar, much like January 1 marks the start of the Gregorian calendar

- The 1st day of the month of Nissan marks the start of the Jewish sacred year

- The 3rd important date is the 9th day of the month of Av which marks the date when Babylon destroyed the Temple at Jerusalem in 586 BC. Other calamitous events have beset the Jewish people on the 9th of Av throughout history. In particular, Cahn tells of the mass expulsion of Jewish people from Spain in 1492. As this expulsion was going on, a certain explorer with 3 ships was about to set sail on a voyage of discovery. That explorer sailed out of port on August 3, 1492 which was one day after the 9th of Av. That explorer was Christopher Columbus and he found the New World and as Cahn tells it, that New World became a new home for the Jewish people

- The 4th key date in the Jewish calendar is Shemini Atzeret (The Gathering of the Eighth Day). This date typically falls somewhere in late September through late October in the month of Tishrei.

The website www.chabad.org allows one to quickly scan back over a number of years to pick off these important dates. I have examined price action on the Dow Jones Average across several years in the context of these key dates. My backtesting has shown a variable correlation to these dates. But, the correlation is intriguing enough that I recommend traders and investors pay attention to these dates.

For 2020, these four key dates fell as follows: 1st of Nissan on March 26, the 9th of Av on July 30, the 1st of Tishrei on September 19 and Shemini Atzeret on October 10.

The 1st of Nissan aligned perfectly to the March 2020 market sell-off lows. Not much happened at the 9th of Av. The 1st of Tishrei marked a swing low on the S&P 500. Shemini Atzeret marked a swing high on the S&P 500. These correlations are further evidence that these dates are critical to watch.

For 2021, these dates will fall as follows: 1st of Nissan on March 14, the 9th of Av on July 18, the 1st of Tishrei on September 7, and Shemini Atzeret on September 28.

A 2005 article in Trader's World magazine suggested that W.D. Gann might have been instilled with knowledge of Jewish mysticism based on the Kabbalah. Gann apparently had connections to a New York personality called Sepharial who is said to have taught Gann about astrology and esoteric matters.

The Kabbalah centers around the Hebrew Alef-bet (alphabet). The Hebrew Alef-bet comprises 22 letters. In Kabbalistic methodology, these letters are assigned a numerical value. Starting with the first letter, values are 1, 2 3 4 5 6 7 8 9 10 20 30 40 50 60 70 80 90 100 200 300 and 400.

There are many mathematical techniques that can be applied to parsing the Alef-bet. One in particular involves taking the odd-numbered letters and the even numbered letters and assigning their appropriate numerical values.

The numerical value (sum total) of the Alef-bet is 1495. The sum total of the odd-numbered letters is 625. The sum total of the even numbered letters is 870.

- 625/1495 = 42%. Taking a circle of 360 degrees, 42% is 150.5 degrees

- 870/1495 = 58%. Taking a circle of 360 degrees, 58% is 209.5 degrees

- Kabbalists are also well aware of phi as it pertains to the Golden Mean. Phi is famously known as 1.618

- 1/phi = 62%. Taking a circle of 360 degrees, 62% is 222.5 degrees

- 1 − (1/phi) = 48%. Taking a circle of 360 degrees, 48% is 137.5 degrees.

From a significant price low (or high), one can examine price charts for time intervals when a geocentric or heliocentric planet advances 137.5, 150.5, 209.5 or 222.5 degree amounts.

To illustrate, consider the significant low in March 2009 on US equity markets (S&P 500) as a start point. Using the phi approach, Figure 25 shows that from the March 2009 lows, 11 years later, 137.5 degree advancements of heliocentric Venus land spot on the March 2020 lows. Figure 25 also shows the second part of the double-top market high in early September 2020 aligned to a Venus 137.5 degree interval.

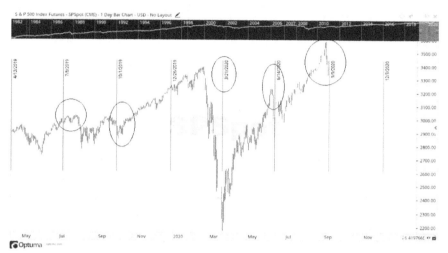

Figure 25
Kabbalistic Math and Venus 137.5 degrees

Looking ahead into 2021, there will be 137.5 degree heliocentric Venus intervals coming into play on the S&P 500 on February 24, May 18, August 11, November 6 and January 31, 2022.

To further illustrate the effect of these intervals, consider the 2011 significant high for Gold. Over eight years later, Figure 26 shows how the 150.5 degree advancements of heliocentric Venus have aligned themselves with uncanny accuracy to various short term trend swings on Gold price. Applying 137.5 degree advancement intervals from the 2011 high shows a perfect alignment to the September 2020 top.

From the September 2020 high on Gold prices, use heliocentric data tables to forecast the 137.5 degree advancements of Venus. January 17, April 13 and July 12 are 2021 dates to get you started.

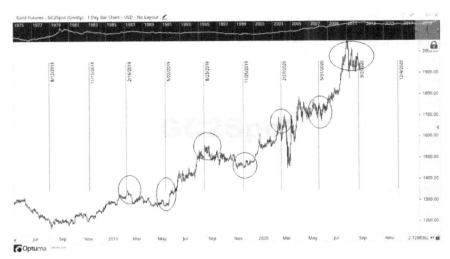

Figure 26
Kabbalistic Math and Gold

Once you source your heliocentric planetary data, either from an on-line site or from a purchased ephemeris book, you can look at practically any commodity or market index for a significant high or low. Apply the 137.5, 150.5, 209.5, or 222.5 degree intervals of Venus forward in time from that start point. You should see a striking pattern emerge. As a final example, from the 2016 multi-year low on Wheat futures the application of the 222.5 degree intervals of Venus yields a very compelling alignment to swing highs and lows. Looking ahead into 2021, other intervals will fall on April 5, August 19, and January 5, 2022.

CHAPTER EIGHT
Synodic, Sidereal & Pythagorean

Astrologers measure time by synodic cycles and sidereal cycles. A sidereal cycle is one that is measured from the vantage point of the Sun. If an observer were standing on the Sun, he or she would see Venus travel around the Sun one complete time in 225 days. Mars would take 687 days. The outer planets would take much longer. In fact, Saturn would take 29.42 years, Uranus 83.75 years, Neptune 163.74 years, and Pluto 245.33 years.

A synodic cycle is measured from the vantage point of here on Earth. To an observer standing here on terra firma, the time it takes Venus to record a conjunction with Sun until that same conjunction occurs again appears to be 584 days. Mars takes 780 days from Sun/Mars conjunction to the next Sun/Mars conjunction. Saturn takes 376 days and the other outer planets 367 to 399 days.

The sidereal cycles of the outer planets bear an alignment to larger events in history. For example, the year 1776 is key to American history. Add a Neptune cycle to 1776 and one gets to 1939 when the world was on the cusp of World War II. The American Civil War started in 1861 with the events at Fort Sumter. Add a Uranus cycle to this date and one gets to the time when World War II ended. Add another Uranus cycle and that takes us to 2028. Are we headed for another major conflict? Current events unfolding around the world would lead one to think so.

The Moon has synodic and sidereal cycles in its orbit around the Earth. The sidereal period of the Moon is 27.5 days and the synodic period is 29.53 days. This latter period lends itself to the expression lunar month. During each sidereal lunar cycle, the Moon can be seen to vary in its position above and below the lunar ecliptic. In a sidereal period, the Moon will go from maximum declination to maximum declination. It is said that W.D. Gann was a proponent of following lunar cycles when trading Soybeans and Cotton. The question is, was he doing so based on the premise that the Moon's gravitational pull governs the ocean tides? With our bodies being substantially water, was he postulating that the Moon was influencing Soybean and Cotton traders? Or, was he part of a larger contingent of traders who were timing the markets to their own advantage at Moon declination highs and lows?

Figure 27 presents a chart of Soybeans with the Moon declination in the lower panel. Note the propensity for declination maxima to align to trend changes. A 4-hour chart is the ideal way to study these price swings closely. You can see why a trader like W.D. Gann would have used this declination phenomenon to give himself an advantage in the markets.

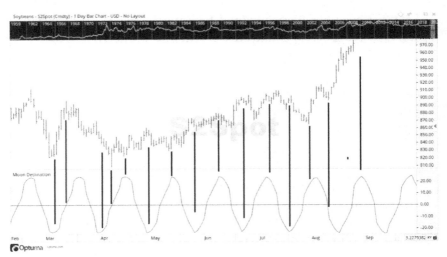

Figure 27
Moon Declination and Soybeans

Figure 28 illustrates Cotton futures prices during 2020 to date. I have overlaid the chart with a dark vertical line at some maximum, minimum and zero point lunar declination events. An argument can easily be made that lunar declination patterns align to short term trend swings on the Cotton market.

If you are a trader of commodity futures, I encourage you to study your favorite futures contracts to see if a Moon declination correlation exists. If it does, take advantage of it.

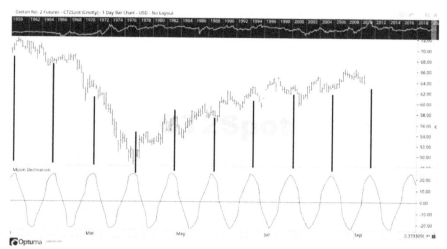

Figure 28
Moon Declination and Cotton

To assist you in some backtesting of your own, consider that in 2020, Moon recorded maximums in its declination on January 10, February 6, March 5, April 1, April 29, May 25, June 22, July 19, August 16, September 12, October 9, November 5, December 4, and December 30.

In 2020, declination minima occurred January 23, February 19, March 18, April 14, May 11, June 7, July 5, August 1, August 29, September 25, October 22, November 18, and December 16.

For 2021, Moon will be at its maximum declination: January 1, January 27, February 23, March 22, April 19, May 16, June 12, July 9, August 6, September 2, September 29, October 27, November 23, December 20.

For 2021, Moon will be at its minimum declination: January 13, February 9, March 8, April 4, May 1, May 29, June 26, July 23, August 20, September 15, October 12, November 9, and December 6.

There are also cyclical intervals that relate to Pythagorean mathematics. Recall from high school math class that the Pythagorean Theorem says for a right angled triangle, $c^2 = a^2 + b^2$. The following mathematical explanations use this theorem to identify cyclical intervals.

Imagine a square with each side being 1000 units long. Now imagine a circle that is inscribed into that square. The diameter of that circle will be 1000 units.

Next, recall from high school mathematics classes the peculiar feature 'pi' which carries the value 1.34159.

The circumference of the circle inscribed into the square will be pi x diameter. In this case pi x 1000 = 3141.159 units.

The perimeter of the square will be the sum total of the four sides, 1000 + 1000 + 1000 + 1000 = 4000.

The ratio of the perimeter to the circumference of the circle is 4000 / 3141.159 = 1.2734. This figure times 1000 equals 1273.4.

Next, imagine the square cut diagonally across. If the sides of the square are each 1000 units, the length of the diagonal is defined by the Pythagorean equation $a^2 + b^2 = c^2$. The diagonal is thus defined by $(1000)^2 + (1000)^2 = c^2 = 1414.21$ units. The ratio of the diagonal to the perimeter of the square is 1414.21/4000 = 0.3536.

Next imagine a square with sides equal to 1273. The diagonal calculates as $(1273)^2 + (1273)^2 = c^2 = 1800.3$ units.

Now multiply 0.3536 by 1800.3 to get 636.6.

Imagine a right angled triangle with one side being 1 unit long and another side being 2 units long. The Pythagorean expression $(1)^2 + (2)^2 = c^2$ can be used to define the length of the longest side (the hypotenuse). In this case, c = 2.2361 units. Now, add the length of all three sides and divide by the length of the longest side (the hypotenuse). This works out to (1 + 2 + 2.2361) / 2.2361 = 2.3416 units.

Consider next the sum of the two sides divided by the hypotenuse. This works out to (2 + 1)/2.2361 = 1.3416.

A circle contains 360 degrees. Another unit of measure for expressing this is radians where 360 degrees equals 2 x pi radians. In our right angled triangle with sides of 1, 2, and 2.2361 units, the angles in that triangle will necessarily be 30, 60, and 90 degrees. An angle of 60 degrees equals pi / 3 radians.

Consider a circle of diameter 1000 units inscribed in a square with sides of 1000 units. The expression (pi x 1000) / 3 = 1047.2 units.

Double this figure and one gets 2094.4 units.

Consider a right-angled triangle with sides of 1047.2 and 2094.4. The Pythagorean theorem says the length of the hypotenuse is $(1047.2)^2 + (2094.4)^2 = c^2$. Solving for c yields 2341.6.

Taking all of the above calculated numbers in these various complicated mathematical expressions and rounding them off slightly gives us the following sequence of numbers: 637, 1047, 1273, 1341, 1414, 1800, 2094, 2341, and 4242.

As I was preparing this manuscript, I took the time to read W.D. Gann's 1927 book entitled *Tunnel Through the Air*. This book is part sci-fi fantasy, part romance story and part war story. Peppered throughout are curious references to times when the main figure Robert Gordon executed trades on Cotton, Wheat and a stock called Major Motors. In one part of the story, Gann draws from the biblical book of Daniel and lists a couple cyclical intervals that Robert Gordon used in trading. The book of Daniel says: *And from the time that the daily sacrifice is taken away, and the abomination of desolation is set up, there shall be one thousand two hundred and ninety days.* The book of Daniel also says: *Blessed is he who waits, and comes to the one thousand three hundred and thirty-five days.* Looking at the above list of Pythagorean related numbers, the figure 1273 is close to the 1290 days referenced in Daniel. The figure 1341 is close to the 1335 referenced in Daniel.

Taking the above list of numbers to be trading days (not calendar days), from a significant low (or high) on a commodity, a stock or an index, one can plot these intervals on the chart. One will often notice a very high propensity for the intervals to align to points of trend change.

To illustrate, consider the price chart of Gold in Figure 29 where I have selected the 2011 price high as a start point. Note how the various intervals align to inflection points on the chart.

If you are wondering where this bit of complex math comes from, credit goes to Bill Erman who passed away in 2016. He called his mathematical approach Ermanometry.

Figure 30 continues the display of Gold intervals. Starting at the price lows of 2015, where I believe the current Gold rally started, a 637 period interval falls just shy of a 2018 swing low. The 1047 period interval falls just shy of the March 2020 COVID-induced spasm.

Curiously, there is a 1272 period interval due very early in 2021. Stay alert for what may become of this interval.

Figure 29
Gold and Ermanometry Intervals

Figure 30
Gold and Ermanometry Intervals

As another example, consider Soybeans and their 2012 extreme price high. Figure 31 presents some of the Ermanometry intervals. Note the curious alignment to points of trend change. The 1414 period interval aligns to a steep plunge in Beans price in 2018. More importantly, there appears to be a 2094 period interval due right near the end of 2020.

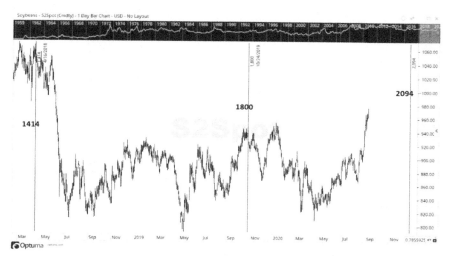

Figure 31
Soybeans and Ermanometry Intervals

Turning to the S&P 500 and applying these intervals starting at a low point in early February 2016, the 1047 trading day interval lands right at the March 2020 lows. Such are the hidden mysteries of geometry.

Look at a chart of interest to you. Identify a significant high or low turning point and then apply the calendar day intervals 637, 1047, 1273, 1341, 1414, 1800, 2094, 2341, and 4242 days. You now have a powerful market timing tool at your fingertips.

CHAPTER NINE
Parallel & Contra Parallel

As the various celestial bodies make their respective journeys around the Sun, they can be seen to move above and below the celestial equator plane. Celestial bodies experience declinations of up to about 25 degrees above and below the celestial equator plane.

Declination can be viewed one planet at a time or by pairs of planets. Let's suppose that at a particular time Mars can be seen as being at 10 degrees of declination above the celestial equator and at that same time Venus is at 9 degrees of declination. Let's further suppose that we allow for up to 1.5 degrees tolerance in our measurement of declinations. We would then say these two planets are at parallel declination. Let's take another example and suppose that at a given time Jupiter is at 5 degrees of declination above the celestial equator. At that same time Pluto is at 6.5 degrees declination below the celestial equator. Again, let's allow for up to 1.5 degrees of tolerance. We would say that Jupiter and Pluto are at contra-parallel declination. Parallel and contra-parallel events can have a powerful bearing on the financial markets.

I have spent many hours exploring parallel and contra-parallel events with respect to commodity futures contracts. This research was spurred on by a 25-year-old astrology book I found in a used bookshop in early 2017. In studying this old book, I have found there are some commodities that bear a correlation to the same parallel and contra-parallel conditions that were in place at the date the contracts first started trading on an exchange (First Trade date). I strongly suggest using these parallel and contra-parallel events in combination with the other astrology events outlined throughout this Almanac. Obtain the planetary declinations at the first trade date for the stock or commodity in question. Examine the data for evidence of parallel and contra parallel events. Watch for these parallel and contra-parallel occurrences to repeat themselves at future dates. To illustrate, consider the following Gold, Cotton, Crude Oil, and Soybeans studies:

Gold

Gold futures started trading in New York on Dec 31, 1974. At that date, the planets were at the following declinations relative to the ecliptic:

Sun	-23.06 degrees
Venus	-22.42 degrees
Mars	-22.45 degrees
Jupiter	-7.39 degrees
Saturn	+22.05 degrees
Uranus	-11.37 degrees
Neptune	-20.3 degrees
Pluto	+11.44 degrees

Looking closer at these numbers we can see the following:

- Uranus is contra-parallel to Pluto (contra-parallel= declinations within 1.5 degrees of each other, but signs are opposite)

- Saturn is contra-parallel Mars

- Saturn is contra-parallel Venus

- Saturn is contra-parallel Sun

- Mars is parallel Sun (parallel = declinations with 1.5 degrees of each other, signs are the same)

- Mars is parallel Venus

- Sun is parallel Venus.

For 2020:

- Mars was parallel Sun for 4 days either side of January 14 and again September 5

- Mars was parallel Venus for about 6 days either side of October 15

- Sun was parallel Venus for 4 days either side of June 4

- Saturn was contra-parallel Venus March 24 through 31

- Saturn was contra-parallel Sun from May 26 through June 10.

The price chart of Gold in Figure 32 has been overlaid with some of these events from 2020. Gold staged a powerful recovery after selling off with the March 2020 COVID panic. This recovery aligns to Saturn being contra-parallel Venus. Saturn being contra-parallel Sun and Sun at the same time being parallel Venus set the stage for a rally that lasted into August 2020.

Figure 32
Gold and Declination Phenomenon

In *Tunnel Through the Air*, Gann references days when the story hero Robert Gordon was very certain that a trend change would occur on his various Major Motors trades. My instinct was to check planetary declination at Major Motors first trade date and compare those figures to Robert Gordon's key dates. Just as I suspected, there is a declination connection. Gann hints strongly that instead of watching for two planets to become parallel or contra-parallel so as to match the situation at the first trade date, one should instead be watching for Mars and Venus in particular to pass the same declination level as they were at on the first trade date. In the case of Gold, August 2018 should have delivered a trend change thanks to Mars being at -22 degrees declination. Mid-February, 2019 should have given a trend change thanks to Venus being at -22 degrees declination. Likewise for the period November-December, 2019 thanks to Venus. In fact, all of these intervals did deliver a minor trend change. Late February-early March, 2020 should have delivered a trend change thanks to Mars being at -22 degrees declination. In fact, that is exactly what happened. Starting February 26, Gold dropped over $100 per ounce. Price then recovered only to get trashed starting March 9 as COVID fears overtook the news headlines. Moreover, Moon had passed its maximum declination on the week-end. More recently, Gold hit a turning point on August 6, 2020. The decination of Saturn at that date was contra-parallel to Saturn's +22 degrees of declination at Gold's first trade date. Moon was at 0 degrees declination. Somehow, Gann liked to tie the declination of Moon into his forecasts and he looked for times of minimum, zero and maximum declination.

The month of January 2021 will have Venus passing through -22 degrees declination. The same will occur in early October, 2021 and again in mid-late December.

Cotton

Cotton futures started trading in New York on June 20, 1870. At that date, the planets were at the following declinations relative to the ecliptic:

Sun	+23.45 degrees
Venus	+14.84 degrees
Mars	+21.42 degrees
Jupiter	+21.32 degrees
Saturn	-22.09 degrees
Uranus	+22.31 degrees
Neptune	+6.86 degrees
Pluto	+2.99 degrees

Looking closer at these numbers we can see the following:

- Sun is contra-parallel to Saturn

- Sun is parallel Uranus

- Saturn is contra-parallel Uranus

- Uranus is parallel Jupiter

- Uranus is parallel Mars

- Mars is parallel Jupiter

- Mars is contra-parallel Saturn

- Mars is parallel Uranus

For 2020:

- Saturn was contra-parallel Sun from May 26 through June 10

- Sun was parallel Uranus for about 6 days either side of April 25.

The price chart of Cotton in Figure 33 has been overlaid with these events from 2020. In each case, a profitable price move resulted.

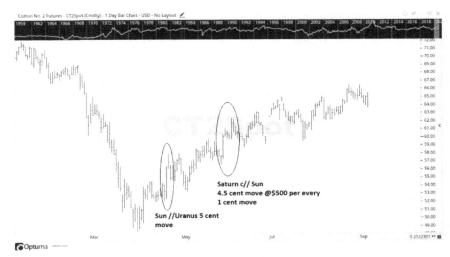

Figure 33
Cotton and Declination Phenomenon

Following the *Tunnel Through the Air* method, in mid-January 2020, Cotton turned in a trend change that would see price fall 24 cents per pound. At ths trend change, Saturn and Jupiter (two heavyweight planets) were parallel in declination to where they had been at Cotton's first trade date. Mars was contra-parallel in declination. Moon had just made its maximum declination. The downward plunge in price came to an end at the end of March 2020 with Moon at maximum declination and Mars and Jupiter parallel to their first trade date declinations.

For 2021, Mars will be parallel to its natal declination in early March and again in early June. Late February and mid-late September will have Venus contra-parallel its natal declination. Late April and mid-July will have Venus parallel its natal declination. March and June will have Mars parallel its natal declination. Jupiter and Saturn will not be factors in 2021.

Crude Oil

Crude Oil futures started trading in America on March 30, 1983. At that date, the planets were at the following declinations relative to the ecliptic:

Sun	+3.52 degrees
Venus	+16.28 degrees
Mercury	+4.38
Mars	+9.52 degrees
Jupiter	-21.18 degrees
Saturn	-9.87 degrees
Uranus	-21.70 degrees
Neptune	-22.20 degrees
Pluto	+5.42 degrees

Looking closer at these numbers reveals the following:

- Sun is parallel to Pluto (declinations within 2 degrees of each other, signs are the same)

- Sun is parallel Mercury

- Mars is contra-parallel Saturn

- Uranus is parallel Jupiter

- Uranus is parallel Neptune

- Jupiter is parallel Neptune

For 2020:

- Sun was parallel Pluto from January 1 through 21.

The price chart of Crude Oil in Figure 34 has been overlaid with this event. Is it just a random co-incidence that this parallel event marked the peak in Oil price?

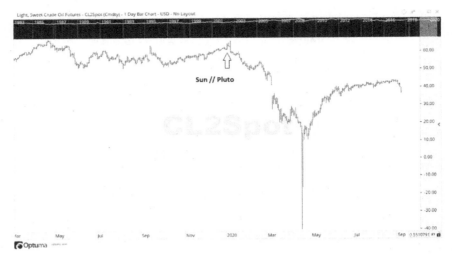

Figure 34
Crude Oil and Declination Phenomenon

Following the *Tunnel Through the Air* methodology, Crude made a trend change starting January 8, 2020 which turned into an ugly thrashing. Venus was contra-parallel, Jupiter parallel and Mercury parallel their respective natal declinations. Moon was at maximum declination. The price beating came to an end April 20 when trend changed. Jupiter was parallel its natal declination at the time. Moon was at zero declination.

For 2021, the month of April could be interesting for Crude. Mercury will move from being contra-parallel to being parallel its natal declination. Late April will see Venus parallel its natal declination. Early August will see Mars parallel its natal declination. Late August and into early September will see Mercury at work again. Early July and late September will see Venus at work. Late October will see Mars at contra-parallel its natal declination.

Soybeans

As will be discussed in a coming chapter, the Chicago Board of Trade was founded April 3, 1848. Looking at Soybean futures through the lens of this date as opposed to the 1936 date when Soybean futures actually started trading yields some interesting finds. At the 1848 date, the planets were at the following declinations relative to the ecliptic:

Sun	+5.35 degrees
Venus	-7.26 degrees
Mercury	-6.10
Mars	+24.80 degrees
Jupiter	+23.26 degrees
Saturn	-6.08 degrees
Uranus	+6.48 degrees
Neptune	-11.48 degrees
Pluto	-5.46 degrees

- Sun is parallel Uranus

- Sun is contra parallel Pluto, Venus, Mercury and Saturn

- Mercury is parallel Saturn

- Mercury is contra parallel Uranus

- Mars is parallel Jupiter

For 2020:

- Sun was parallel Uranus from April 21 to 27

- Sun was contra parallel Saturn from May 14 through June 4

- Sun was contra parallel Pluto from May 18 through June 17

- Mars was parallel Jupiter from March 10 through April 15.

The price chart of Soybeans in Figure 35 has been overlaid with some of these events from 2020.

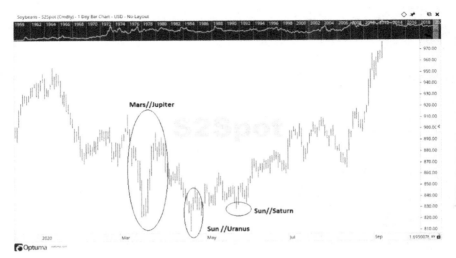

Figure 35

Soybeans (1848) and Declination Phenomenon

The Mars parallel Jupiter event in March saw Beans go on a wild excursion. A significant low in April aligned to the Sun parallel Uranus event. A minor dip lower in May aligned to the Sun parallel Saturn event.

In the context of *Tunnel Through the Air*, January 9, 2020 saw trend change on Soybeans that resulted in price declining almost $1.50 per bushel by April 21. The January trennd change likely was triggered by Moon at maximum declination, Jupiter contra-parallel its natal declination and Mars within 3 degrees contra-parallel its natal declination. The April low saw Moon at zero declination and Jupiter contra parallel its natal declination. As I pen these words, I note that Soybeans are looking rather over-extended. As part of the recent rally, there was a short, sharp pullback in price from September 18 through 30. Jupiter was contra-parallel its natal declination and Moon went from zero declination down to minimum and back to zero. I suspect in November to see Soybeans ease off their uptrend with Venus approaching its contra-parallel declination and Jupiter still hovering around its contra-parallel level.

For 2021, early March and early September will see Venus parallel its natal declination. Early April and early August will see a contra-parallel situation. The April-May period will see Mars parallel its natal declination.

Of all the astrology techniques that I have used, this one involving parallel and contra-parallel is the most tedious. This is where a software program such as Solar Fire Gold can help you with its built-in tables of data and its visual display of declination pattern for various planets.

Once you determine the stock or commodity you wish to follow, you can use Solar Fire Gold to determine what declination situations existed at the first trade date (the date when the stock or commodity first began trading on an exchange). Then for any calendar year in question, you can see when these declination events repeat themselves.

CHAPTER TEN
NYSE 2021 Astrology

No examination of the astrology of the New York Stock Exchange would be complete without mention of Louise McWhirter. After years of reading old papers and manuscripts, I still have no idea who Louise McWhirter was. What I do know is during her lifetime, Louise McWhirter focused intently on the astrology of the New York Exchange. Her technique which revolves around the New Moon (lunation) remains viable to this day.

The Lunation and the New York Stock Exchange

A lunation is the astrological term for a New Moon. At a lunation, the Sun and Moon are separated by 0 degrees which means the Sun and Moon are together in the same sign of the zodiac. The correlation between the monthly lunation event and New York Stock Exchange price movements was first popularized in 1937 by McWhirter. In her book, *Theory of Stock Market Forecasting*, she discussed how a lunation making hard aspects to planets such as Mars, Jupiter, Saturn and Uranus was indicative of a coming month of volatility on the New York Stock Exchange. She also paid close attention to Mars and Neptune, the two planets that rule the New York Stock Exchange. The concept of planetary rulership extends back into the 1800s with each zodiac sign having a planet that rules that sign. McWhirter said those times of a lunar month when the transiting Moon makes 0 degree aspects to

Mars and Neptune should be watched carefully. McWhirter arrived at her Mars and Neptune rulership conclusion by observing that the 10th House of the 1792 birth horoscope wheel for the NYSE spans Pisces and Aries. Neptune rules Pisces and Mars rules Aries.

New York Stock Exchange – First Trade Chart

The New York Stock Exchange officially opened for business on May 17, 1792. As the horoscope in Figure 36 shows, the NYSE has its Ascendant (Asc) at 14 degrees Cancer and its Mid-Heaven (MC) at 24 Pisces.

McWhirter further paid close attention to those times in the monthly lunar cycle when the transiting Moon passed by the NYSE natal Asc and MC locations at 14 Cancer and 24 Pisces respectively.

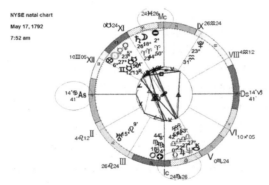

Figure 36
NYSE First Trade horoscope

Horoscope Charts and the McWhirter Methodology

In my research and writing, I follow the McWhirter methodology for shorter term trend changes. When forecasting whether or not a coming month will be volatile or not for the NYSE, the McWhirter methodology starts with creating a horoscope chart for the New Moon date and positioning the Ascendant of the chart at 14 degrees Cancer (the Ascendant position on the 1792 natal chart of the New York Stock Exchange). Positioning the Ascendant is made easy in the Solar Fire Gold software program. Aspects to the lunation are then studied.

If the lunation is at a 0, 90, or 120 degree aspect to Mars, Neptune, 14 Cancer, or 24 Pisces, one can expect a volatile month ahead. A lack of such aspects portends a less volatile period. The McWhirter method demands a consideration of where the Moon is at each day. Aspects of the Moon to Mars, Neptune, 14 Cancer, or 24 Pisces represent dates of potential short term trend reversals. Although not expressly stated by McWhirter, it is also important to pay attention to those dates when Moon is at either maximum or minimum declination. As well, dates when Mercury is retrograde and dates when Venus is at or near its maximum or minimum declination should be considered carefully.

Similarly, when studying an individual stock or an individual commodity futures contract, the McWhirter approach calls for the creation of a horoscope chart at the First Trade date of the stock or commodity. The Ascendant is then shifted so that the Sun is at the Ascendant. The software program Solar Fire Gold is very good for generating First Trade horoscope charts for McWhirter analysis where the Ascendant needs to be shifted.

In stock and commodity analyses, McWhirter paid strict attention to those times of a calendar year when transiting Sun, Mars, Jupiter, Saturn, Neptune, and Uranus made hard 0, 90, and 180 degree aspects to the natal Mid-Heaven, natal Ascendant, natal Sun, natal Jupiter and even the natal Moon of the individual stock or commodity future being studied.

One must be alert at these aspects for the possibility of a trend change, the possibility of increased volatility within a trend, or even the possibility of a breakout from a chart consolidation pattern. Evidence of such trend changes will be found by watching price action relative to moving averages and by utilizing oscillator type functions (MAC-D, DMI, RSI and so on).

McWhirter Lunation Past Example

The McWhirter method can best be illustrated by examining the S&P 500 chart for 2020 year to date and overlaying it with key points.

In Figure 37, a point labelled Cancer refers to Moon transiting past the 14 of Cancer point, MH or Pisces refer to Moon transiting past 24 of Pisces, and points labelled Mars or Neptune refer to Moon passing Mars or Neptune (co-ruling planets of the NYSE).

Figure 37
McWhirter events 2020 year to date

As can be seen in Figure 37, Moon passing Neptune and the 24 Pisces point certainly contributed along with other astro phenomena to the start of the COVID panic sell-off. Moon passing Neptune helped to define the March lows. Since the March lows, the flood of fiat liquidity into the system has been over-whelming. The various points labelled on the chart since March have helped to define short, sharp swings in price. It is however interesting to see that Moon passing Neptune and the 24 Pisces natal mid-Heaven point align to the start of the September sell-off.

In past editions of my annual Almanac publications, I presented images of the horoscope wheel at each New Moon event. I then went on to list a number of events from each month's lunar cycle. In this edition, I dispense with the overly complicated treatment. What follows in this chapter is a listing of the date for each lunar cycle in 2021 along with a list of times when Moon passes Mars, Neptune, the NYSE natal Mid-Heaven at 24 Pisces, and the NYSE natal Ascendant at 14 Cancer.

2021 Lunation Events

December 2020 to January 2021

Market action from late December through late January 2021 will be influenced by the New Moon cycle that commences on December 14, 2020 with Sun at 23 Sagittarius. NYSE co-ruler Neptune is positioned within 6 degrees of the NYSE natal mid-Heaven point. Neptune is also positioned at the apex of a right angled triangle pattern. The lunation itself is 90 degrees square to the NYSE natal mid-Heaven point. Heliocentric Jupiter is now a couple degrees separated from Saturn as the Gann Master Cycle begins anew. This lunar cycle runs until January 12, 2021 and could well be an energized one. It is important to note that there is a geocentric version of the Gann Master Cycle as well. Geocentric Jupiter and Saturn will be exactly conjunct on December 14, 2020.

Key dates during this lunar cycle are:

- December 14: geocentric Jupiter and Saturn are conjunct

- December 20-21: Moon passes Neptune and natal mid-Heaven point

- December 23: Moon passes co-ruler Mars

- December 28 - 31: Moon at maximum declination as it passes 14 of Cancer

- Early January: Venus records its declination low.

January 2021 to February 2021

The January New Moon cycle commences on January 12, 2021 with Sun at 22 Capricorn. This lunar cycle runs until February 11, 2021. The lunation itself is a favorable 60 degrees to the NYSE natal mid-Heaven point. Venus is 120 degrees trine to Mars.

Key dates during this lunar cycle are:

- January 13: Moon at declination minimum

- January 17: Moon passes Neptune and natal mid-Heaven point

- January 20: Moon passes co-ruler Mars

- January 23: Mercury at greatest easterly elongation

- January 26: Moon passes 14 of Cancer

- January 27: Moon at maximum declination

- January 30: Mercury turns retrograde

- February 9: Moon at minimum declination.

February 2021 to March 2021

The February New Moon cycle commences on February 11, 2021 with Sun at 23 Aquarius. This lunar cycle runs until March 13, 2021. The lunation itself is within orb of being square to NYSE co-ruler Mars.

Key dates during this lunar cycle are:

- February 13: Moon passes Neptune and natal mid-Heaven point

- February 18: Moon passes co-ruler Mars

- February 19: Mercury retrograde completes. It is likely that a trend change will occur during the Mercury retrograde event. Attention is warranted

- February 23: Moon passes 14 of Cancer and makes maximum declination

- March 6: Mercury at greatest westerly elongation

- March 8: Moon at declination minimum.

March 2021 to April 2021

The March New Moon cycle commences on March 13, 2021 with Sun at 24 Pisces. This lunar cycle runs until April 12, 2021. The lunation itself is conjunct the NYSE natal Mid-Heaven and conjunct NYSE co-ruler Neptune. This lunar cycle could thus be energized.

Key dates during this lunar cycle are:

- March 19: Moon passes co-ruler Mars

- March 22: Moon passes 14 of Cancer and hits declination maximum

- March 26: Venus at Superior Conjunction. Such conjunction events can lead to trend changes. Caution is warranted.

- April 4: Moon at declination low

- April 9: Moon passes co-ruler Neptune and natal Mid-Heaven at 24 Pisces.

April 2021 to May 2021

The April New Moon cycle commences on April 12, 2021 with Sun at 22 Aries. This lunar cycle runs until May 11, 2021. The lunation itself is not in any hard aspect to other planets. Mars is a favorable trine 120 degrees to Jupiter.

Key dates during this lunar cycle are:

- April 17: Moon passes co-ruler Mars

- April 19: Moon passes 14 of Cancer and makes its

- declination maximum

- April 22: Mars at its declination maximum

- May 1: Moon at declination low

- May 6-7: Moon passes co-ruler Neptune and natal Mid Heaven at 24 Pisces.

May 2021 to June 2021

The May New Moon cycle commences on May 11, 2021 with Sun at 20 Taurus. This lunar cycle runs until June 10, 2021. The lunation itself is at a 90 degree hard aspect to Saturn. This combined with the fact that Mars is parked atop the natal NYSE Ascendant at 14 Cancer and Neptune is atop the NYSE natal Mid Heaven suggests an energized cycle.

Key dates during this lunar cycle are:

- May 16: Moon passes co-ruler Mars and the 14 of Cancer point and makes declination maximum

- May 17: Mercury at greatest easterly elongation

- May 29: Moon at declination low

- May 31: Mercury turns retrograde

- June 3: Moon passes natal Mid Heaven and co-ruler Neptune

- Early June: Venus begins the process of making its declination maximum

June 2021 to July 2021

The June New Moon cycle commences on June 10, 2021 with Sun at 19 Gemini. This lunar cycle runs until July 10, 2021. The lunation itself is at within orb of being 120 degree trine Saturn and 90 degrees square Mars. This suggests an energized cycle.

Key dates during this lunar cycle are:

- June 12: Moon passes 14 of Cancer and makes a declination maximum.

- June 13: Moon passes co-ruler Mars

- June 21: Mercury retrograde complete

- June 26: Moon at minimum declination

- June 30: Moon passes co-ruler Neptune and natal Mid Heaven

- June 30: Moon passes co-ruler Neptune and natal Mid Heaven at 24 Pisces

- July 4: Mercury at greatest westerly elongation

- July 9: Moon at maximum declination

July 2021 to August 2021

The July New Moon cycle commences on July 10, 2021 with Sun at 18 Cancer. This lunar cycle runs until August 8, 2021. The lunation itself is at a 120 degree trine aspect to the natal Mid-Heaven point.

Key dates during this lunar cycle are:

- July 12: Moon passes co-ruler Mars

- July 23: Moon at declination low

- July 28: Moon passes co-ruler Neptune and natal Mid Heaven

- August 5: Moon passes 14 of Cancer and makes a declination maximum

August 2021 to September 2021

The August New Moon cycle commences on August 8, 2021 with Sun at 15 Leo. This lunar cycle runs until September 7, 2021. The lunation is at a 90 degree square to Uranus.

Key dates during this lunar cycle are:

- August 10: Moon passes co-ruler Mars

- August 20: Moon at declination low

- August 24: Moon passes co-ruler Neptune and the natal Mid-Heaven

- September 2: Moon passes 14 of Cancer and makes declination maximum.

September 2021 to October 2021

The September New Moon cycle commences on September 7, 2021 with Sun at 14 Virgo. This lunar cycle runs until October 5, 2021. The lunation itself is at a 120 degree trine to Uranus.

Key dates during this lunar cycle are:

- September 8: Moon passes co-ruler Mars

- September 13: Mercury at greatest easterly elongation

- September 15: Moon at declination low

- September 21: Moon passes co-ruler Neptune and the natal Mid-Heaven

- September 28: Mercury turns retrograde. A trend change is very likely in Mercury retrograde events

- September 30: Moon passes 14 of Cancer and makes its declination maximum.

October 2021 to November 2021

The October New Moon cycle commences on October 5, 2021 with Sun at 13 Libra. This lunar cycle runs until November 4, 2021. The lunation itself sits right atop NYSE
co-ruler Mars and is a hard 90 degrees to the 14 of Cancer point. This portends a lunar cycle full of energetic activity.

Key dates during this lunar cycle are:

- October 6: Moon passes co-ruler Mars

- October 12: Moon at declination low

- October 16: Mercury retrograde completes

- October 17: Moon passes co-ruler Neptune and the natal Mid-Heaven at 24 Pisces

- October 25: Mercury at greatest westerly elongation

- October 26: Moon passes 14 of Cancer point and completes its declination maximum.

November 2021 to December 2021

The November New Moon cycle commences on November 4, 2021 with Sun at 12 Scorpio. This lunar cycle runs until December 4, 2021. The lunation itself sits at a favorable 120 degree trine to the 14 of Cancer point.

Key dates during this lunar cycle are:

- Early November: Venus at declination minimum

- November 9: Moon at declination low

- November 13: Moon passes co-ruler Neptune and the natal Mid-Heaven

- November 22: Moon passes the 14 of Cancer point and completes its declination high

- December 2: Moon passes co-ruler Mars.

December 2021 to January 2022

The December New Moon cycle commences on December 4, 2021 with Sun at 13 Sagittarius. This lunar cycle runs until January 2, 2022. The lunation itself is not in detrimental aspect to any planets.

Key dates during this lunar cycle are:

- December 6: Moon at declination low

- December 11: Moon passes co-ruler Neptune and natal Mid Heaven point

- Mid-December: Venus turns retrograde. Trend changes are highly likely during Venus retrograde events. Caution is warranted.

- December 20: Moon passes 14 of Cancer point and makes a declination maximum

- December 31: Moon passes co-ruler Mars.

CHAPTER ELEVEN

Commodities 2021 Astrology

Gold

Investors who own Gold are accustomed to routinely checking the price of Gold by tuning into a television business channel or perhaps obtaining a live quote of Gold futures. What many do not realize is that Gold is a unique entity. For quietly working behind the scenes is an archaic methodology called the London Gold Fix.

The London Gold Fix occurs at 10:30 am and 3:00 pm local time each business day in London. Participants in the daily fixes are: Barclay's, HSBC, Scotia Mocatta (a division of Scotia Bank of Canada) and Societe Generale. These twice daily collaborations (some would say collusions) provide a benchmark price that is then used around the globe to settle and mark-to-market all the various Gold-related derivative contracts in existence.

The history of the Gold Fix is a fascinating one. On the 12th of September 1919, the Bank of England made arrangements with N.M. Rothschild & Sons for the formation of a Gold market in which there would be one official price for Gold quoted on any one day. At 11:00 am, the first Gold fixing took place, with the five principal gold bullion traders and refiners of the day present. These traders and refiners were N.M. Rothschild & Sons, Mocatta & Goldsmid, Pixley & Abell, Samuel Montagu & Co. and Sharps Wilkins.

The horoscope in Figure 38 depicts planetary positions at this date in history. Observations that jump off the page include: North Node had just changed signs, Venus was retrograde, Sun and Venus were conjunct, Mercury and Saturn were conjunct, Mars, Neptune and Jupiter were all conjunct at/near the Mid-Heaven point of the horoscope, and Saturn was 180 degrees opposite Uranus.

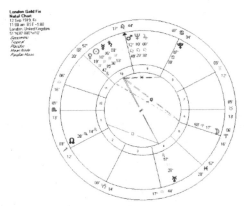

Figure 38
1919 London Gold Fix horoscope

Gold investors who have been around for a while will remember the significant $800/ounce price peak recorded by Gold in January 1980. To illustrate how astrology is linked to Gold prices, consider that at this price peak the transiting North Node had just changed signs and was 90 degrees hard aspect to the natal Node in the 1919 horoscope. Consider too that Mars and Jupiter were both coming into a 0 degree conjunction with the natal Sun location in the 1919 horoscope. For those who were involved in Gold more recently, recall that Gold hit a significant peak in early September 2011 at just over $1900/ounce. At that peak, Sun and Venus were conjunct to one another as they were in the 1919 Gold Fix horoscope. What's more, they were within a few degrees of being conjunct to the natal Sun location in the 1919 horoscope. A coincidence ? I say not.

In the few weeks that followed this 2011 peak, Gold prices plunged nearly $400/ounce. But, then Gold found its legs again and began to rally. This rally seems directly related to Mars coming into a 0 degree conjunction to the Mars-Jupiter-Neptune location of the 1919 horoscope wheel.

Such is the complex nature of Gold prices. I have studied past charts of Gold and I am shocked at how many price inflection points are related in one way or another to the astrology of the 1919 Gold Fix horoscope wheel. To those readers who are of the opinion that Gold price is manipulated, your notion is indeed a valid one. I believe that astrology is the secret language being spoken amongst those that play a hand in the manipulation.

Gold futures contracts started trading in America on the New York Mercantile Exchange on December 31, 1974. Figure 39 illustrates the planetary positions in 1974 at the first trade date of Gold futures.

Note that in the 1919 chart Mars and Neptune are conjunct one another. Now, observe in Figure 39 that Mars and Neptune are also conjunct in the 1974 chart.

Next, ask yourself why the New York Mercantile Exchange would launch a new futures contract on December 31, a time when most staff would be off for Christmas holidays. If this seems more than a bit odd, you are not alone in your thinking.

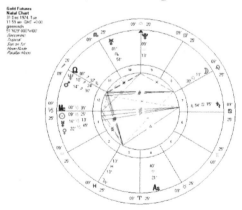

Figure 39
Gold futures First Trade horoscope

Note the location of Moon in the 1974 horoscope at 11 degrees of Leo. Now, look at the 1919 horoscope and observe that 11 degrees of Leo is where Mars and Neptune are located. I take these curious placements as further evidence of a deliberately timed astrological connection between Gold price, the 1919 Gold Fix date, and the 1974 first trade date for Gold futures. All very intriguing stuff to be sure.

Transiting Mars/1974 natal Sun

Times when certain celestial objects transit past key points in the 1974 first trade horoscope deserve attention. One transiting body to consider is Mars. Transiting Mars passing conjunct (0 degrees) to the natal Sun location from the 1974 First Trade horoscope is a valuable tool for Gold traders to consider. The chart in Figure 40 has been overlaid with the two most recent conjunct aspects of transiting Mars / natal Sun. In early 2018 the Mars/natal Sun aspect sparked a trend change and Gold price fell by nearly $200 in the months that followed. The next conjunct aspect just so happened to fall right at the beginning of the March 2020 COVID-induced Gold sell-off. A co-incidence, you say?

Figure 40
Transiting Mars / 1974 natal Sun conjunction

In 2021, transiting Mars will not make a conjunct aspect to the natal Sun location. However, early 2022 will see such a conjunction.

Transiting Sun and Mars /1919 natal Sun

Times when certain celestial objects transit past key points in the 1919 Gold fix horoscope also warrant attention. For 2021, transiting Sun will make 0 degree conjunct aspects to the 1919 natal Sun location:

- September 2-September 19: passing 0 degrees conjunct to 1919 natal Sun

For 2021:

- Transiting Mars will not make any conjunct aspects to the 1919 natal Sun location.

Sun Conjunct Venus

Another cue from the 1919 chart is the conjunction between Sun and Venus. Gold prices recorded a trend change in early September 2019 when Venus and Sun were conjunct from July 24 through about September 22. This trend change inflection point remained unchallenged for the following 4 months. In 2021 from mid-February through April, Sun and Venus will again be passing through conjunction.

Mercury Retrograde

Another valuable tool for Gold traders to consider is Mercury retrograde events. Watch for technical chart trend indicators to suggest a short term trend change at a retrograde event.

The chart in Figure 41 illustrates the connection between these Mercury phenomena and Gold prices from late 2019 to this time of writing in September 2020. In late 2019, a Mercury retrograde event marked a price low and the start of an uptrend. The March 2020 plunge in Gold prices was immediately preceded by the end of a Mercury retrograde event. The retrograde event in July 2020 had no effect. In other words, the power players on Wall Street reserve the right to use Mercury retrograde events to their advantage as they see fit. All we can do as traders and investors is to be alert for possible trend changes at the Mercury retrograde events.

For 2021, Mercury will be:

- retrograde from January 30 through February 19

- retrograde from May 31 through June 21

- etrograde from September 28 through October 16.

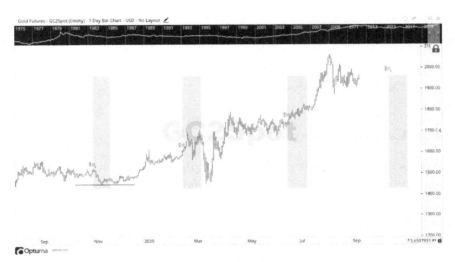

Figure 41
Mercury retrograde and Gold prices

Reminder - Follow the Trend

At the risk of sounding overly repetitive, I must stress again the importance of following the trend. A question that I routinely get is when during one of these astrological transit events should a person implement a trade? The answer is very simple. You should consider implementing a trade when you see the trend change. Always let the trend be your friend. I am sure you have heard this mantra before.

There are many ways of measuring trend. My experience has shown me that the methodologies developed by J. Welles Wilder are very powerful for identifying trend changes. In particular I prefer to use his Wilder Volatility Stop. Wilder's 1978 book, New Concepts in Technical Trading Systems, is a highly recommended read if you are seeking to learn more about his methods. Another technique for gauging trend is to overlay a price chart with moving averages, such as the 34-day and 55-day ones. By the way, these are Fibonacci numbers and I routinely use them in my personal examination of trend. Lately I have also been studying the Ichimouku Cloud methodology as a trend system.

Silver

Silver futures started trading on a recognized financial exchange in July 1933. Figure 42 shows the First Trade horoscope for Silver futures in geocentric format.

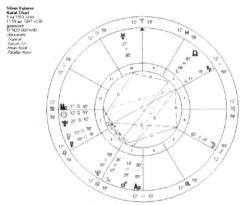

Figure 42
Silver futures First Trade horoscope

My research has shown that times when transiting Sun, transiting Mars and transiting Jupiter make hard aspects to the natal Sun point at 12 degrees Cancer should be watched carefully for evidence of trend changes and price inflection points. I am intrigued with this First Trade date. I suppose Silver could have started trading anytime in 1933. July 4, 1776 is a critical date in US history and on this date Sun was at 14 Cancer. Recall that 14 Cancer also figures prominently in the first trade horoscope of the NYSE. In 1933, markets would have been closed for the 4th of July celebrations. A first trade date of July 5, is as close as authorities could come to the July 4 date. On July 5, Sun at 12 degrees, 59 minutes is within a degree of the critical 14 of Cancer point.

Jupiter/natal Sun

In April 2011, Silver prices reached a peak at just under $50 per ounce. Transiting Jupiter was making a 90 degree aspect to natal Sun at the time. From this peak, Silver prices declined towards a significant low in late 2015. Along the way, transiting Jupiter made a 0 degree conjunction to natal Sun in the August 2013 timeframe. Silver prices behaved extremely erratically during this period. During October, November and December 2016, Jupiter made a 90 degree hard aspect to the natal Sun position. Silver price moved quickly from $17 to $19/oz. and then fell hard down to $15.50.

In early 2020, Silver prices again were impacted by Jupiter passing 180 degrees opposite to the 1933 natal Sun location. As Figure 43 illustrates, the end of this transit was marked by the start of a sharp decline in Silver prices, which just so happened to coincide with the COVID market sell-off. So, was it COVID that caused Silver to sell off? Or was it market players lurking in the shadows manipulating markets in accordance with astrological cycles?

Figure 43
Silver futures and Jupiter / natal Sun

As I pen this manuscript, I see that Jupiter is again making another 180 degree opposition aspect to the natal Sun. This second aspect comes about because Jupiter turned retrograde after the March 2020 event, only to turn direct again in August 2020. As you read this page, you can look back at a chart of Silver and see what, if any, effect this transit had on prices.

The next hard aspect of Jupiter to natal Sun (a 90 degree square) will occur in mid-2022.

Sun/natal Sun

In past years, Silver prices have recorded swing highs and lows at times when transiting Sun made 0, 90 and 180 degree aspects to the natal Sun position at 12 Cancer. Evidence for such reactions in 2020 to date has been sparse. But, these aspects should not be ruled out going forward. For 2021, Sun will make aspects to natal Sun as follows:

- March 25-April 7: Sun 90 degrees to natal Sun

- June 26-July 10: Sun will pass 0 degrees to natal Sun

- September 28-October 10: Sun passes 90 degrees to natal Sun.

Mars/natal Sun

The Silver price chart in Figure 44 has been overlaid with times when transiting Mars makes 0, 90, and 180 degree aspects to the natal Sun position. Note the tight correlation to the March 2020 price plunge. At this time, Mars was passing 180 degrees opposite the natal Sun. In July 2020, Silver prices exploded higher, perhaps triggered by a Mars square aspect to natal Sun.

Figure 44

Silver and Mars/natal Sun events

For 2021, Mars will make aspects to natal Sun as follows:

- May 4 - May 23: Mars will pass 0 degrees to natal Sun.

- September 24-October 13: Mars will pass 90 degrees to natal Sun.

Declination

Planetary declinations should also be considered when studying price action of Silver futures. In particular the declination maxima and minima of Venus and also of Sun should be watched. Why Venus? As it turns out, Venus had just made its maximum declination in 1933 as Silver futures were starting to trade for the very first time. This is a very strong hint that Venus declination should be examined relative to Silver price inflections. In 2020, Silver prices began to rally in earnest following the early May Venus maximum declination event. This rally was given added impetus in June with Sun making its declination maxima.

- For 2021, Sun will be at its maximum declination at the Summer Solstice on June 21. Sun will at its minimum declination at the Winter Solstice on December 21

- For 2021, Venus will exhibit its maximum declination from mid-May through mid-June.

Copper

The First Trade Date for Copper futures was July 29, 1988. Figure 45 illustrates the First Trade horoscope in geocentric format.

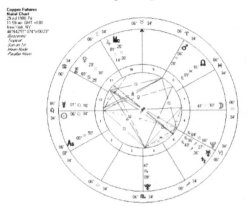

Figure 45
Copper futures First Trade horoscope

This horoscope wheel features an Inferior Conjunction of Mercury in the sign of Leo. Also, notice in this horoscope that the First Trade date is that of a Full Moon.

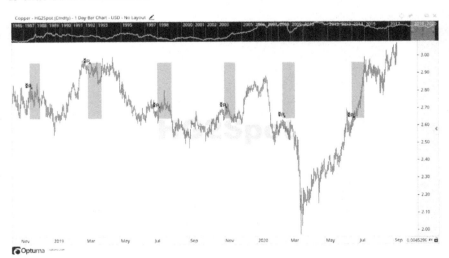

Figure 46
Copper and Mercury retrograde events

An Inferior Conjunction of Mercury marks the start of a new Mercury cycle around the Sun. Mercury Inferior Conjunction events always occur in association with Mercury being retrograde.

Figure 46 shows Copper prices overlaid with Mercury retrograde events. Knowing that retrograde events are approaching, one should watch for short term trend changes.

For 2021, Mercury will be:

- retrograde from January 30 through February 19

- retrograde from May 31 through June 21

- retrograde from September 28 through October 16.

Canadian Dollar, British Pound & Japanese Yen

These three futures instruments all started trading on May 16th, 1972 at the Chicago Mercantile Exchange. The horoscope in Figure 47 illustrates planetary placements at this date. It is interesting to note that Mars is 180 degrees opposite Jupiter. This suggests that Mars and Jupiter may play a role in price fluctuations on these currencies. Mars is also 0 degrees conjunct to Venus, suggesting another cyclical relationship.

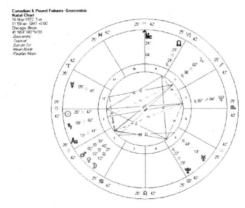

Figure 47
Pound, Yen, Canadian First Trade horoscope

The Mars-Venus Influence

In the natal horoscope in Figure 47, notice that Mars and Venus are conjunct. Mars conjunct Venus events only occur every couple years. A Mars/Venus conjunction event occurred most recently in August 2019. For the latter half of 2019, the Canadian Dollar traded in a sideways pattern. The swing low recorded in August at the Mars/Venus conjunction proved to be the low (which was tested twice) for the entire latter half of 2019. The next Mars/Venus conjunction will occur in July 2021.

Natal Transits

Transiting Sun passing natal Sun, natal Mars and quite often natal Jupiter are events that currency traders may wish to focus on.

To illustrate, the chart in Figure 48 illustrates the effect on the British Pound of transiting Sun passing natal Mars (2 degrees of Cancer in the 1972 horoscope chart) and Sun passing natal Sun (25 of Taurus in the 1972 horoscope). In the May through July timeframe of 2020, a Sun natal Sun conjunction and a Mars/natal Sun conjunction both aligned to sharp V-bottoms.

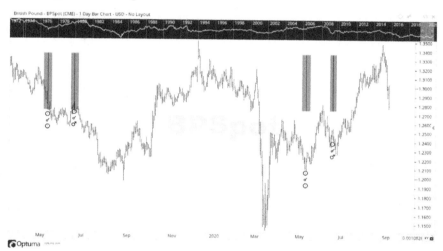

Figure 48
British Pound Sun passing natal Mars & natal Sun

The chart in Figure 49 illustrates the effect on the Canadian Dollar of transiting Sun passing natal Jupiter. Note the alignment to a swing top in 2019 and a swing low in 2020.

For 2021, transiting Sun will make hard aspects to the natal Mars point of 2 degrees Cancer as follows:

- June 16 to June 29: 0 degrees conjunct

- For 2021, transiting Sun will make hard aspects to the natal Sun point of 25 Taurus as follows:

- May 8 to May 24: 0 degrees conjunct

- For 2020, transiting Sun will make hard aspects to the natal Jupiter point of 7 degrees Capricorn as follows:

- December 21 to January 4, 2022: 0 degrees conjunct.

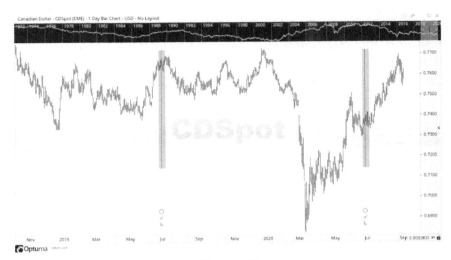

Figure 49
Canadian Dollar Sun conjunct natal Jupiter

Mercury Retrograde

Currency traders should pay close attention to Mercury retrograde events as they can bear a good alignment to trend changes on the Pound, Yen and Canadian Dollar.

The Canadian Dollar price chart in Figure 50 has been overlaid with Mercury retrograde events. All too often there is a correlation to swing highs and lows.

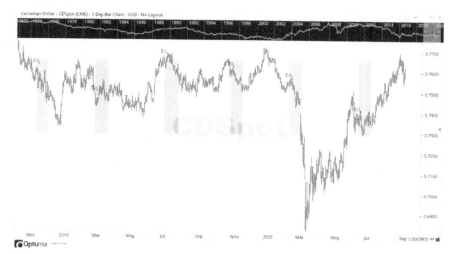

Figure 50
Canadian Dollar and Mercury retrograde

For 2021, Mercury will be:

- retrograde from January 30 through February 19

- retrograde from May 31 through June 21

- retrograde from September 28 through October 16.

Euro Currency Futures

The Euro became the official currency for the European Union on January 1, 2002 when Euro bank notes became freely and widely circulated.

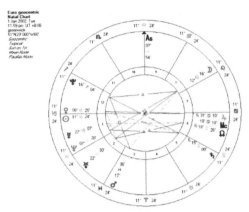

Figure 51
Euro Currency First Trade horoscope

Natal Transits

Events of transiting Sun making 0, and 90 degree aspects to the natal Sun position in the Euro 2002 First Trade horoscope are also worth watching as they often align to inflection points on the Euro. In 2019 notice how these transits all came within close proximity to price inflection points.

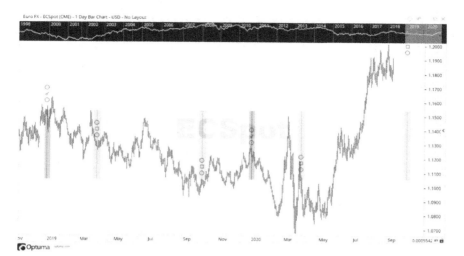

Figure 52
Natal Transits and the Euro Currency

For 2021, transiting Sun will be:

- At a 0 degree conjunction to natal Sun during the first week of January

- 90 degrees to natal Sun March 26 through April 7

- 90 degrees to natal Sun from 28 of September through October 9.

Australian Dollar

Australian dollar futures started trading on the Chicago Mercantile Exchange on January 13, 1987. As the horoscope in Figure 53 shows, Sun and Mercury are at Superior Conjunction at 22-23 degrees Capricorn.

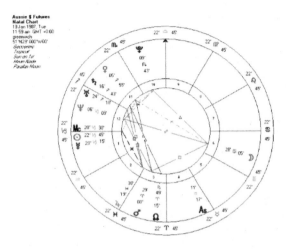

Figure 53
First Trade horoscope of Australian Dollar futures

Mercury Retrograde

Times when Mercury is retrograde should be considered when trading Australian Dollar futures. The chart in Figure 54 has been overlaid with Mercury retrograde events. These events often align to price inflection points.

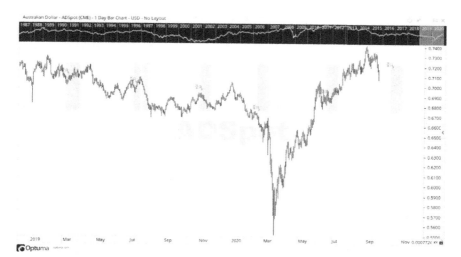

Figure 54
Australian Dollar and Mercury Retrograde

For 2021, Mercury will be:

- retrograde from January 30 through February 19

- retrograde from May 31 through June 21

- retrograde from September 28 through October 16.

Natal Transits

Transits to the natal Sun position of 22 Capricorn from the 1987 First Trade horoscope can be used as tools to help one navigate the price action of the Australian Dollar as the price chart in Figure 55 illustrates.

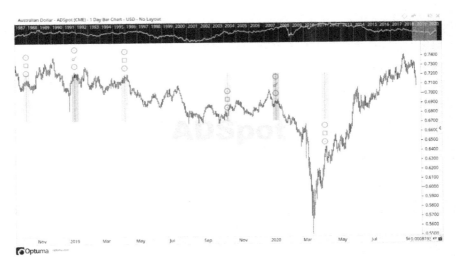

Figure 55

Australian Dollar and Sun/natal Sun events

For 2020, transiting Sun will make 0 and 90 degree aspects to natal Sun:

- 0 degrees to natal Sun from January 5 through January 21

- 90 degrees square to natal Sun from April 6-18th.

- 90 degrees to natal Sun from October 9-21st .

30-Year Bond Futures

30-Year Bond futures started trading in Chicago on August 22, 1977. Figure 56 presents the geocentric First Trade horoscope for this date.

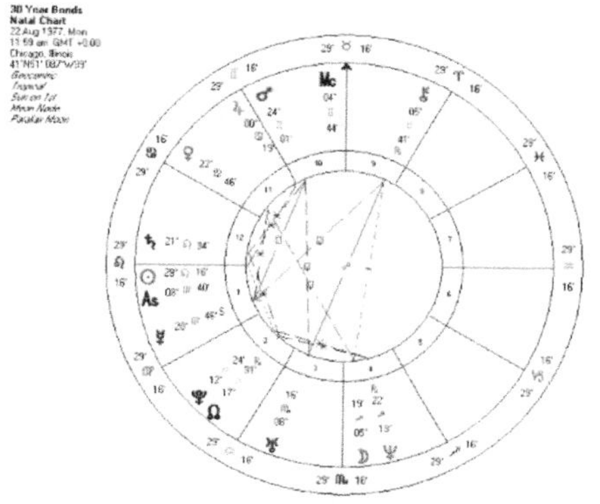

Figure 56
First Trade horoscope for 30-Year Bond futures

Natal Transits

My research has indicated that events of transiting Mars making 0 and 90 degree aspects to the natal Mars position at 24 degrees Gemini are valuable tools for the Bond trader. And by the way, in the 1776 natal horoscope of the USA, Mars just so happens to be at 21 Gemini. Figure 57 illustrates Bond price performance with the Mars/natal Mars transits overlaid.

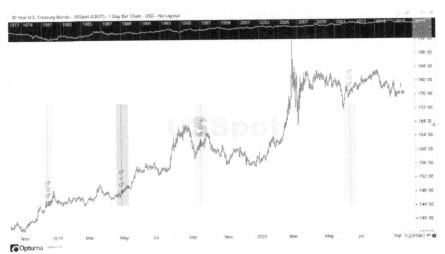

Figure 57
30-Year Bonds and Mars in aspect to natal Mars

For 2021, transiting Mars will make hard 0 and 90 degree aspects to natal Mars as follows:

- April 7 through 18, Mars will pass 0 degrees to natal Mars

- August 31 to September 13, transiting Mars Sun will make a 90 degree aspect to natal Mars.

Mercury Retrograde

Look closely at the First Trade horoscope in Figure 56 and you will note that the position of Mercury (at 20 Virgo) is further delineated by a letter S. This letter denotes stationary and curiously enough this First Trade date of August 22, 1977 comes one day prior to Mercury turning retrograde. Therein rests a strong hint. The price chart in Figure 57 has been overlaid with Mercury retrograde events. Note the propensity for short term inflections in trend at these retrograde events.

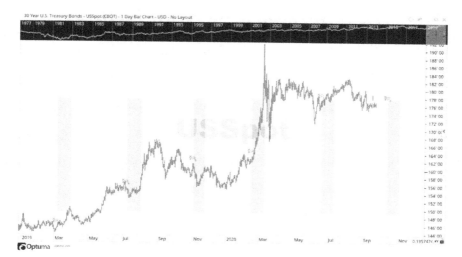

Figure 57
30-Year Bonds and Mercury retrograde

For 2021, Mercury will be:

- retrograde from January 30 through February 19

- retrograde from May 31 through June 21

- retrograde from September 28 through October 16.

10-Year Treasury Note Futures

10-Year Treasury Notes started trading in Chicago on May 3, 1982. Figure 58 presents the geocentric First Trade horoscope for this date.

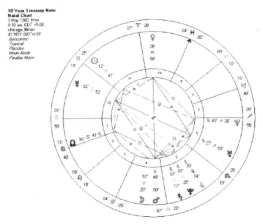

Figure 58
First Trade horoscope for 10-Year Treasury Notes

Retrograde

Notice in the First Trade horoscope in Figure 58, that Mars is denoted Rx which stands for retrograde. Therein rests another valuable clue. If I look back at past Treasury Note performance, I see very significant price inflections at Mars retrograde events. As you read this section, note that Mars was retrograde from early September through to mid-November 2020. As I write this section, I rather suspect that there will be notable volatility as this retrograde event which overlaps with the US Presidential election.

Mercury retrograde events also bear watching when following price action on the 10-Year Treasury Notes. Figure 59 illustrates the connection between price inflection points and Mercury retrograde. Most recently, the retrograde event in July marked a sharp push higher as a flight to safety took place following China being labelled a currency manipulator.

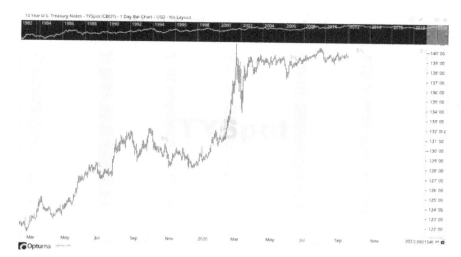

Figure 59
Mercury retrograde and 10-Year Treasuries

For 2021, Mercury will be:

- retrograde from January 30 through February 19

- retrograde from May 31 through June 21

- retrograde from September 28 through October 16.

Wheat, Corn, Oats

Wheat, Corn and Oats futures all share the same first trade date from 1877. The horoscope in Figure 60 shows planetary placements at this date.

Figure 60
First Trade horoscope for Wheat, Corn and Oats futures

Over the past couple years, my reading and research has revealed that W.D. Gann was also known to follow a First Trade horoscope wheel from April 3, 1848, the date the Chicago Board of Trade was founded. Figure 61 shows this horoscope wheel.

In the 1877 Wheat/Corn natal horoscope, the Sun is at 12 Capricorn, exactly square to the location of Sun in the above shown 1848 horoscope.

In the 1877 chart, the Descendant is at 12 Cancer. Look where Jupiter is in this 1848 chart – within 0.5 degrees of 12 Cancer. Co-incidence ? I rather doubt it.

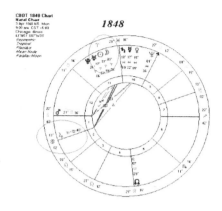

Figure 61
1848 First Trade horoscope for CBOT

Natal Aspects to Both Horoscopes

Events of transiting Sun making 0, 90 and 180 degree aspects to the natal Sun position in the 1877 First Trade horoscope or the 1848 CBOT natal horoscope can be used as a tool to guide traders through the shorter term price volatility of Wheat, Corn and Oats. Because of the peculiar alignment of these two horoscopes, a 0 degree conjunction to natal Sun in the 1877 horoscope will be a 90 degree square to the natal Sun of the 1848 horoscope.

A more important tool is to consider 0 and 90 degree aspects of heliocentric transiting Jupiter to natal heliocentric Jupiter. In early 2016, a 90 degree event saw Corn prices rally almost $1 per bushel, which is $5000 on a single futures contract. A conjunct event in early 2019 saw a similar move. In early 2022, a 90 degree event will again occur. The effect on Wheat prices is also apparent. In 2016, the 90 degree event saw a 50 cent per bushel drop in price at the transit. The conclusion of the 2019 transit saw an explosive $1.30 per bushel rally on Wheat prices.

Mars is also a planet to keep tabs on. In March 2020, Mars passing conjunct to the 1877 natal Sun location triggered a steep fall in Corn prices. In August 2020, a 90 degree square event sparked a rally in Corn prices which at this time of writing is still intact.

For 2021, transiting Mars will be 90 degree to the 1877 natal Sun from September 28 to October 8.

Declination

A sizeable price rally on Corn got underway in May 2019 as Mars made its maximum declination. Mars at a declination low in early 2020 aligned to a price trend change. On the Wheat market, a Mars declination low in early 2020 aligned to a sharp rally which subsequently failed. In November 2019, an 80 cent per bushel rally aligned to Venus making its declination low. In the April-May period of 2020, a Venus declination high aligned to a steady decline in Wheat prices. The Venus declination high in the April-May period of 2020 aligned to a low and a trend change on Corn.

For 2021, Venus will exhibit its maximum declination from mid-May through mid-June. In early November, Venus will be recording its minimum declination. Mars will make its declination maxima in mid-April 2021.

Retrograde

Mercury retrograde plays a role in price pivot points on the grains. The price chart of Wheat futures in Figure 62 has been overlaid with Mercury retrograde events. Although not shown here, Mercury retrograde events do have a similar propensity to align to trend changes on Corn prices.

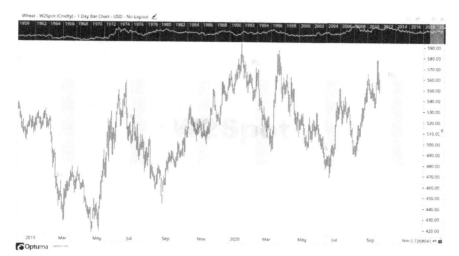

Figure 62
Wheat prices and Mercury retrograde events

For 2021, Mercury will be:

- retrograde from January 30 through February 19

- retrograde from May 31 through June 21

- retrograde from September 28 through October 16.

Soybeans

Soybean futures started trading in Chicago on October 5, 1936. The horoscope in Figure 63 illustrates the planetary placements at that time. What is intriguing is the location of the Sun. Notice how it is exactly 90 degrees to the location of the Sun in the First Trade horoscope for Wheat, Corn and Oats? Notice Sun is 180 degrees from the Sun in the 1848 CBOT natal chart? As I have previously suggested, the regulatory officials who determined these First Trade dates knew more about astrology than we may think.

Figure 63
Soybeans First Trade horoscope

Take a moment to look back at the 1877 natal horoscope for Corn and Wheat. Can you spot any similarity to the 1936 Soybeans horoscope? Hint: look at Jupiter. Are these first trade dates purely random?

Natal Transits

Events of transiting Sun making 0, 90, and 180 degree aspects to the natal Sun position in the 1936 First Trade horoscope can be used to navigate the volatility of the Soybean market. Figure 64 illustrates the effect of transiting Sun making 0 and 90 degree aspects to the natal Sun position. Figure 65 illustrates the effect of Mars making aspects to the natal Sun location.

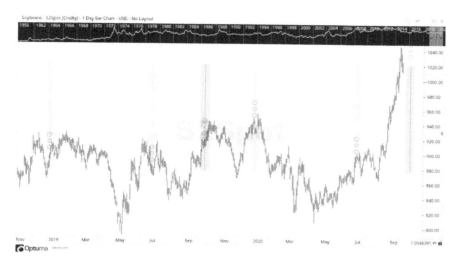

Figure 64
Soybeans and Sun/natal Sun events

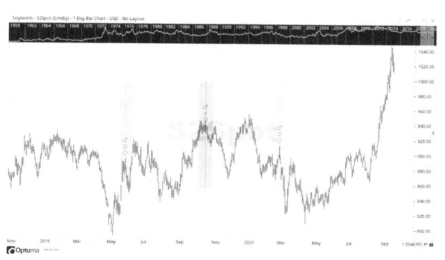

Figure 65
Soybeans and Mars/natal Sun events

Following on with the Jupiter position, in 2016 heliocentric Jupiter making a 90 degree aspect to the natal heliocentric Jupiter position saw Beans plunge by $2.50 per bushel. In early August 2019, heliocentric Jupiter passing 0 degrees to the natal heliocentric Jupiter location saw a $1 per bushel rally unfold. A 90 degree transit will next occur in 2022.

For 2021:

- Transiting Sun will make a 90 degree aspect to natal Sun from January 1 through January 8

- A 90 degree aspect to natal Sun will occur from June 28 through July 11

- A 0 degree aspect will occur from September 26 through October 12.

For 2021:

- Mars will be 90 degrees to the 1936 natal Sun May 4th through May 22nd

- Mars will be 0 degrees to the 1936 natal Sun from September 26 through October 11.

Retrograde

Mercury retrograde events also contribute to the price behavior of Soybeans. The Soybeans chart in Figure 66 illustrates the Mercury retrograde effect. If there is a trend change associated with Mercury retrograde, the trend change may come immediately beforehand, during or immediately afterwards. The use of a suitable chart technical indicator is essential to help identify the trend shifts.

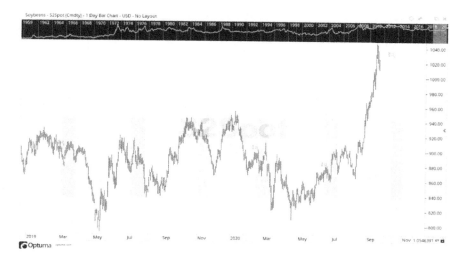

Figure 66
Soybeans and Mercury events

For 2021, Mercury will be:

- retrograde from January 30 through February 19

- retrograde from May 31 through June 21

- retrograde from September 28 through October 16.

Declination

Soybeans also have a tendency to record price trend changes in proximity to Venus recording maximum, and minimum declinations. The Soybean price chart in Figure 67 illustrates further. A price high and trend turning point in early 2019 saw a sizeable price decline. A declination maxima in mid-2019 fell in between a double top formation. A declination low in late 2019 aligned to a swing low and trend change. A declination maxima in 2020 aligned to what was arguably the start of a sizeable rally.

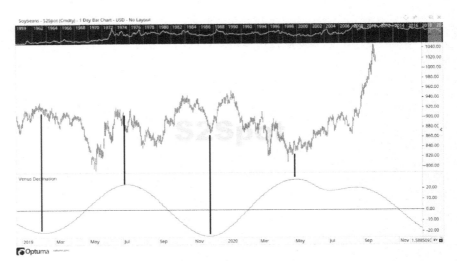

Figure 67
Soybeans and Venus Declination

For 2021, Venus will exhibit its maximum declination from mid-May through mid-June. In early November, Venus will be recording its minimum declination.

Crude Oil

West Texas Intermediate Crude Oil futures started trading on a recognized exchange for the first time on March 30, 1983. A unique alignment of celestial points can be seen in the horoscope in Figure 68. Notice how Mars, North Node, (Saturn/Pluto/Moon) and Neptune conspire to form a rectangle.

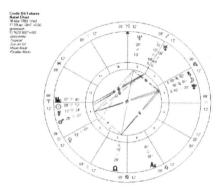

Figure 68
Crude Oil First Trade horoscope

Natal Transits

My experience has shown that Crude Oil is a complex instrument to analyze using astrology. Given the peculiar rectangular shape that appears in the horoscope, my strategy for analyzing Crude Oil has been to use natal transits with a focus on transiting Sun and transiting Mars making 0 degree aspects to the four corner points of the rectangle.

The chart in Figure 69 illustrates Oil price action with events of Mars transiting the corners of the peculiar horoscope rectangle. In May 2019, Mars passing the Node corner of the rectangle resulted in a substantive $14 drawdown in price. Mars passing the Neptune corner of the rectangle in February 2020 firmly pushed price momentum to the downside.

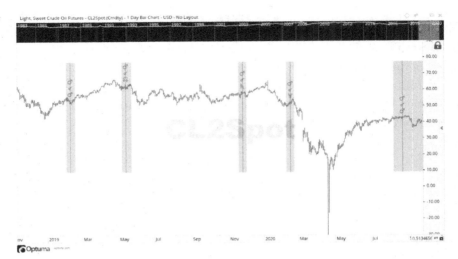

Figure 69
Crude Oil and Mars transits

Figure 70 illustrates events of Sun passing the rectangle corner positions in 2019 to date. A key reversal in price at the $42 level came in December 2018 as Sun passes the Neptune corner of the rectangle. A price high and trend change came in April 2019 at $66 as Sun passed the Mars corner of the rectangle. A price low at $51 in June came as Sun passed the Node corner of the rectangle. The alignment of these Sun transits to swings and trend changes is remarkable.

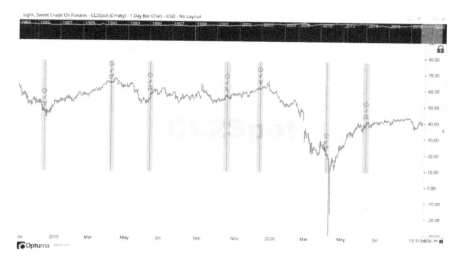

Figure 70
Crude Oil and Sun transits

For 2021, the four corners of the peculiar rectangle will be passed by as follows:

- Sun will transit 0 degrees to the natal Mars location from April 9 through April 22

- Sun will transit 0 degrees to the natal Node location from June 7 through June 22

- Sun will transit 0 degrees to the natal (Saturn/Pluto/Moon) location from October 18 through November 1

- Sun will transit 0 degrees to the natal Neptune location from December 14 through the end of the year

- Mars will transit past the natal Mars location between the 13th of December 2020 and the 8th of January 2021

- Mars will transit past the natal Node location between April 15th and May 3rd

- Mars will transit pass the Saturn/Pluto/Moon location between October 21 and November 10.

Transits of heliocentric Jupiter to the natal heliocentric Jupiter location in the 1983 first trade horoscope are also deserving of attention. In 2015, Oil prices fell over $20 per barrel as heliocentric Jupiter passed 90 degrees to the natal heliocentric Jupiter point. Oil prices fell over $30 per barrel in late 2018 as heliocentric Jupiter passed 0 degrees to the natal heliocentric Jupiter point. Late 2021 will see the next 90 degree transit.

Retrograde

Crude Oil is influenced by Mercury retrograde and Venus retrograde.

For 2021, Mercury will be:

- retrograde from January 30 through February 19

- retrograde from May 31 through June 21

- retrograde from September 28 through October 16.

For 2021, Venus will be retrograde from mid-December through to late January 2022.

Cotton

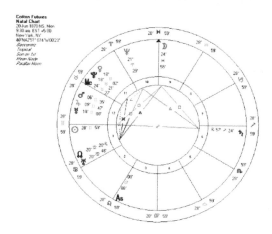

Figure 71
Cotton futures First Trade horoscope

After sifting through back-editions of New York newspapers, I have come to conclude that Cotton futures first started trading on June 20, 1870. The horoscope wheel in Figure 71 illustrates planetary placements at that time. At first glance, I find it peculiar that the Moon is at the same degree and sign location (24 Pisces) as is the Mid-Heaven in the New York Stock Exchange natal horoscope wheel from 1792. Mars, Jupiter and Mercury also are clustered around the same location (9 Gemini) where one finds Uranus in the USA 1776 natal chart. Surely the selection of this Cotton first trade date is no accident.

Natal Transits

Events of transiting Sun passing 0 and 90 degrees to the natal Sun position are an effective tool for traders to use when navigating the choppy waters of Cotton prices. The Cotton price chart in Figure 72 illustrates further.

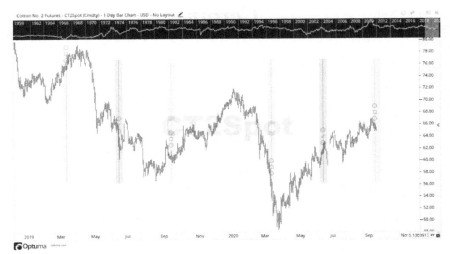

Figure 72
Cotton futures and Sun/natal Sun aspects

In 2021, transiting Sun will aspect natal Sun as follows:

- Transiting Sun will pass 90 degrees to natal Sun from March 7 through March 23

- Transiting Sun will pass 0 degrees conjunct to natal Sun from June 10 through June 30

- Transiting Sun will pass 90 degrees to natal Sun from September 13 through September 30

- Transiting Sun will pass 180 degrees to natal Sun from December 12 through December 28.

Figure 73
Cotton futures and Venus/natal Moon aspects

One other astro phenomenon traders may wish to consider as a tool to use is the occurrence of Venus passing by the natal Moon position at 24 Pisces.

Not a frequent event, it is nonetheless one to pay attention to. The price chart in Figure 73 illustrates.

In April 2019, a rally failed at a Venus natal Moon transit event. Cotton prices subsequently declined over 20 cents per pound.

During 2020, Venus transited past the natal Moon from late January through early February. This passage triggered a 20 cent per pound decline in cotton prices. A 3 cent move on Cotton futures is $1500. Do the math and ask yourself if trading Cotton futures is worthwhile. For 2021, Venus will pass the 24 Pisces point from March 12 through March 21.

Coffee

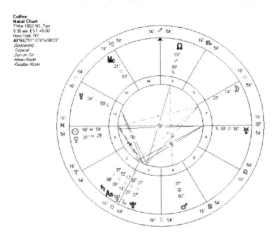

Figure 74
Coffee futures First Trade horoscope

Coffee futures started trading in New York in early March of 1882. The horoscope wheel in Figure 74 illustrates planetary placements at that time.

Natal Transits

In the Coffee horoscope, note the 180 degree aspect between Sun and Uranus. Louise McWhirter in her 1937 writings cautioned it is not wise to invest in situations where this sort of aspect exists because one will experience many wild ups and downs in price over time. A quick look at a 10-year price chart of Coffee reveals a price range of $0.65/pound to $3.06/pound with many wild swings. Your point is well taken, Ms. McWhirter.

The Coffee price chart in Figure 75 has been overlaid with events of transiting Sun making 0 and 90 degree aspects to the natal Sun position at 16 Pisces.

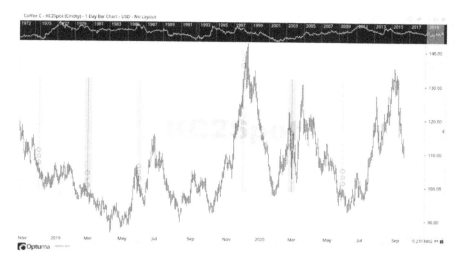

Figure 75
Coffee prices and natal transits

In 2021, transiting Sun will aspect natal Sun as follows:

- Transiting Sun will pass 0 degrees conjunct to natal Sun from February 29 through March 16

- Transiting Sun will pass 90 degrees conjunct to natal Sun from May 31 through June 17

- Transiting Sun will pass 180 degrees to natal Sun from August 31 through September 19

- Transiting Sun will pass 90 degrees to natal Sun from December 10 through December 16.

Non-natal Transits

The positioning of Sun opposite Uranus in the 1882 natal horoscope is intriguing. 0, 90, and 180 degree aspects between transiting Sun and Uranus can be used to further assist the Coffee trader with decision making.

In early 2019, a 90 degree aspect aligned to a trend change. The downward action came to an end at a conjunction event. A 90 degree event in mid-2019 came 7 days ahead of a swing low. A similar such occurrence came again late October.

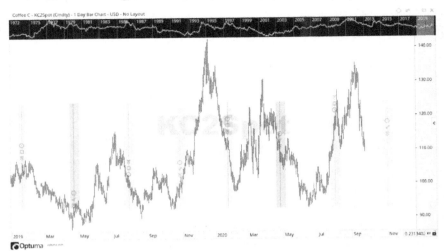

Figure 77
Coffee prices and Sun/Uranus aspects

In late January 2020 a swing low developed within days of a 90 degree aspect. A sharp blow-off rally occurred in late July at a 90 degree aspect.

These aspects between Sun and Uranus will sometimes fall just shy of the actual price turning point, but they do seem to be ominous nonetheless. Figure 77 illustrates recent aspects.

For 2021, Sun will make the following hard aspects (0, 90, 180 degree) with Uranus:

- 90 degrees in late January

- 0 degrees in late April

- 90 degrees in late July

- 180 degrees in late October.

Sugar

Figure 78
Sugar Futures First Trade horoscope

Sugar as a bulk commodity started trading in New York as early as 1881. Old editions of New York newspapers suggest that in September 1914 there were plans to open a formal Sugar Exchange, but these plans were scuttled by World War 1. Following the War, a formal Exchange did open on August 31, 1916. The horoscope wheel in Figure 78 illustrates planetary placements at that time. What stands out on this chart wheel is the T-Square formation with Mars at its apex.

Natal Transits

Events of transiting heliocentric Mars making 0 and 90 degree aspects to the natal heliocentric Mars location at 22 Scorpio have a high propensity to align to pivot swing points. The price chart in Figure 79 illustrates further. This is the intriguing part of astrology. Even though a significant feature might be noted on a geocentric horoscope, sometimes it is a heliocentric event that triggers action. In Figure 79, the significant price peak and reversal in February 2020 came right after a heliocentric Mars/natal Mars conjunction.

Figure 79
Sugar prices and natal transits

For 2021, transiting heliocentric Mars will make the following aspects to heliocentric natal Mars:

- 90 degrees in early to mid June

- 0 degrees in late December.

Retrograde

Mercury retrograde events have a propensity to align to short term trend changes on Sugar price as the chart in Figure 80 illustrates. Keep your eye on Mercury retrograde events and Sugar trends.

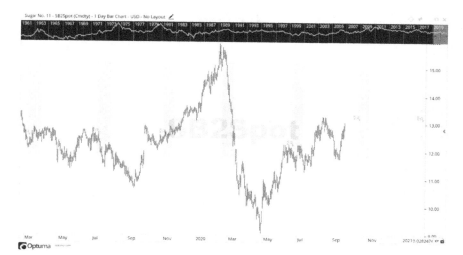

Figure 80
Mercury retrograde and Sugar price

For 2021, Mercury will be:

- retrograde from January 30 through February 19

- retrograde from May 31 through June 21

- retrograde from September 28 through October 16.

Cocoa

Figure 81
Cocoa futures First Trade horoscope

Cocoa futures started trading in New York in early 1925. The horoscope in Figure 81 shows planetary placements at the first trade date. What I find peculiar on this horoscope wheel is the Mid-Heaven point is located at 14 of Cancer, that same mysterious point that appears in the First Trade horoscope of the New York Stock Exchange.

Retrograde

Mercury retrograde events have a high propensity to align to pivot swing points on Cocoa price. The price chart in Figure 82 illustrates further. Sometimes a Mercury retrograde event can deliver some erratic volatility as part of an ongoing trend, while other times the retrograde event can bring about a complete change of trend either immediately before the retrograde or immediately after. Either way, Mercury retrograde events deserve close scrutiny.

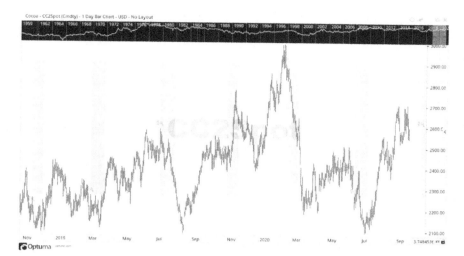

Figure 82
Mercury retrograde and Cocoa price

For 2021, Mercury will be:

- retrograde from January 30 through February 19

- retrograde from May 31 through June 21

- retrograde from September 28 through October 16.

Conjunctions and Elongations

The 1925 natal horoscope shows Sun and Mercury conjunct (0 degrees apart). My research has shown that events of Mercury being at its maximum easterly and westerly elongations and events of Mercury being at its Inferior and Superior conjunctions align quite well to pivot swing points. The price chart in Figure 83 illustrates both phenomena further. I am particularly intrigued with the East and west elongation alignment to pivot price points.

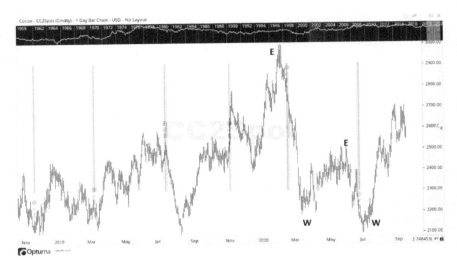

Figure 83
Cocoa price and Mercury cycles

For 2020, Mercury was at greatest easterly elongation February 10, June 3, and October 1. Greatest westerly elongation was at March 24, July 22, and November 10.

For 2021, Mercury will be at its greatest easterly elongation January 23, May 17, and September 13. Greatest westerly elongation events will occur March 6, July 4, and October 25.

CHAPTER TWELVE
Gann, Astrology & Weather

Tunnel Through the Air

In 1927, W.D. Gann wrote a peculiar book entitled *Tunnel Through the Air*. It is a curious story about a figure named Robert Gordon and his life journey that saw him evolve into a successful commodity futures trader. Riddled throughout the book are dates when Robert Gordon placed trades on Wheat, Corn and Cotton, expecting to make money because of pending weather patterns he seems to know about. When the reader actually generates a horoscope for these various trade dates, aspects between some of the larger outer planets are plainly visible in the zodiac wheel. From this, we can infer that aspects between these outer planets can affect the weather and thereby commodity prices.

Uranus and Neptune

Taking cues from the book, it appears that times when Uranus and Neptune make aspects to one another, particularly in January through April, foretell a wet weather period ahead. Wet weather can delay crop planting dates which can impact crop yields and thereby price.

Using Solar Fire Gold software, I did a quick search for times when these two planets were at any geocentric aspect. What I found was a

45 degree aspect during the first 8 months of 2018. For most of this timeframe, Uranus was in the sign of Pisces, a water sign.

I also found a 45 degree aspect in the first 5 months of 2019. For most of this timeframe, Uranus was also in the sign of Pisces.

https://www.nass.usda.gov/Publications/Todays_Reports/ reports/cropan19.pdf is the link to the National Agricultural Statistics Service. This site notes that in 2018, hurricanes Florence and Michael brought destructive damage to coastal areas and also left a footprint inland which affected crops. Verbiage reads: *In many areas, ample rainfall accompanied the above-normal temperatures, leading to record-high yield expectations for the Nation's corn and soybeans. As the growing season ended, rainfall intensified in many areas, contributing to substantial harvest delays. Fieldwork further slowed in some areas in November, when cold, stormy weather hampered final harvest efforts and slowed winter wheat planting, emergence, and establishment across portions of the Plains, Midwest, South, and East.*

A look at charts of Soybeans, Cotton, Wheat, Corn and Cotton shows: Soybeans rose $1 per bushel between September 2018 and December 2018, Cotton prices declined in the same period, Wheat prices trended sideways and Corn prices rose 50 cents per bushel.

https://www.nass.usda.gov/Publications/Todays_Reports/ reports/cropan20.pdf is the link to the 2019 crop overview. Verbiage on the site notes:

Despite the mid-year demise of El Niño, wet conditions were a hallmark of the 2019 growing season across large sections of the Plains and Midwest. Midwestern wetness, which intensified in March and persisted through spring and into summer, caused significant planting delays and subsequently slowed the development of crops such as corn and soybeans. Planting and developmental delays extended to other crops, including sugarbeets, sunflowers, and spring wheat.

A look at charts of Soybeans, Cotton, Wheat, Corn and Cotton shows: Soybeans rose $1.20 per bushel between May 2019 and July 2019, Cotton prices declined in the same period, Wheat prices spiked about $1.30 per bushel and Corn prices spiked over $1 per bushel.

It seems fair to conclude that Uranus/Neptune aspects do have an effect on weather patterns.

If using these aspects as a trading tool, one must watch the price trend for evidence of a technical pattern breakout. In this era of climate change, it can no longer be assumed that these aspects will automatically deliver excess moisture. Moreover, even if moisture does result, it may not be adverse over an entire growing region. As well, crops today have been genetically modified so that even with a planting delay or even with a bit too much moisture, a decent harvest can still result. Farmers today also tend to hold back some product in the bin after a harvest. So what might have given Gann some notable price surges in 1927, may not deliver the same today.

For the first 2 months of 2021, Uranus and Neptune will be at a 45 degree aspect. Keep your umbrella close at hand and watch for adverse weather in the Spring of 2021. Watch the price charts of Wheat, Corn, Soybeans and Cotton closely.

Jupiter and Neptune

Taking further cues from Gann's book, it appears that times when Jupiter and Neptune make aspects to one another, particularly in April through June, foretell a desirable weather period ahead. Desirable growing weather can result in above forecast yields and lower prices come harvest time.

Using Solar Fire Gold software, I did a quick search for times when these two planets were at any geocentric aspect. What I found was a 120 degree aspect in the February-April period of 2014, a 135 degree aspect in the May-June of 2014, a 180 degree aspect in the March-July period of 2016, a 120 degree aspect in the April-September period of 2018, a 90 degree aspect in the 2019 growing season and a 60 degree aspect in the 2020 growing season.

Crop statistics show that weather patterns were not uniform in 2014, but on balance soil moisture conditions did improve over the year. Data for 2016 shows yields on the year were up from 2015. The 120 degree aspect in 2018 conflicts with the Uranus/Neptune aspect of 2018. A positive planetary influence overlain on a negative influence might explain why price increases on futures contracts were not bigger. The 90 degree aspect in 2019 conflicts with the Uranus/Neptune aspect of 2019. It appears that the Uranus/Neptune duo won the weather battle.

It is too soon to say what the 2020 crop looked like under the influence of the Jupiter/Neptune aspect. But, an interim September report is pointing to higher crop yields than in 2019. This should bode poorly for prices heading into the end of 2020. As I write this chapter, grain prices seem to be reaching a blow-off peak.

The Spring of 2022 should be a very good time for planting and growing under the influence of a Jupiter/Neptune conjunction in 'wet' planet Pisces. Watch the price charts closely.

Saturn and Neptune

Taking another cue from the book, it appears that times when Saturn and Neptune make aspects to one another, can promote dry weather which will have a positive impact on prices come harvest time.

For 2021, during the July to November period, Saturn and Neptune will make a 45 degree aspect. Neptune will be in 'wet' sign Pisces, but Saturn will be in 'dry' sign Aquarius. Keep a close watch on price trend for any concerns being expressed over soil moisture conditions that may trigger price surges.

CHAPTER THIRTEEN
Price, Time, & Quantum

Gann Fan Lines

Gann lines are a technique in which a starting point of a significant high or low is selected. From this point, angles (vectors) are projected outwards. These vectors are the 1x1, 1x2, 1x4, 1x8, 2x1, 4x1 and 8x1. In and of themselves, these Gann Lines are not related to astrology. However, in my opinion, they should be applied to charts and used in combination with astrology.

Many market data software platforms will come with a Gann Fan function already built in. The confusion with Gann lines comes from the mathematical method of constructing the lines. In fact, in the Optuma/Market Analyst program there are no fewer than ten ways to apply Gann Lines to a chart. If Mr. Gann were around today he would probably shake his head in bewilderment at how convoluted his technique has become. My preference for applying Gann Lines is the methodology used by Daniel Ferrera in his book *Gann for the Active Trader*. Ferrera's method is based on the Gann Square of Nine mathematics.

To illustrate the creation of Gann lines, I will use Gold which is very much a hot topic these days.

On August 16, 2018 Gold made a price low at $1175. This is the point from which I wish to extend Gann lines.

Step 1: Take the $1175 figure and express it as the number 1175. Take the square root of 1175 and you get 34.27. This will be your time factor.

Step 2: Add 1 to 34.27 and re-square this figure to get 1244.

Step 3: We can now state that our time factor is 34.27 calendar days. For simplicity, we can round this off to 34 days. We can further state that our price factor is 1244 minus 1175 = $69.

Step 4: From the August 16 date extend a line so that it passes through the time co-ordinate (August 16+34 days = September 20) and the price co-ordinate $1244 ($1175 + $69 = $1244). This line is the Gann 1x1 line.

Step 5: From the August 16 date, extend another line so that it passes through the time co-ordinate (August 16 + (34 x 2) days = October 24) and the price co-ordinate $1244. This line is the Gann 1x2 line.

Step 6: From the August 16 date, extend another line so that it passes through time co-ordinate ((August 16+ (4 x 34) days = January 3 and the price co-ordinate $1244. This line is the Gann 1x4 line.

The Gold price chart in Figure 84 has these Gann lines overlaid starting from the August $1175 low.

Figure 84
Gann Lines applied to a Gold Chart

This chart has been prepared in the Optuma/Market Analyst software platform, but you do not necessarily need a fancy software program. A pencil and a ruler could be used to draw lines onto a chart printout.

Notice from the $1175 low point, price action has been constrained by the 1x2 and 1x4 lines. At this time of writing, price action has fallen back below the 1x2 line.

Let's take a look at one more example, using Wheat as the subject. Wheat registered a significant low on April 30, 2019 at $4.16 per bushel.

Step 1: Take the $4.16 value and express it as the number 4160. Take the square root of 4160 and you get 64.49. This will be the time factor.

Step 2: Add 1 to 64.49 and re-square this figure to get 4290. Shifting the decimal over we get $4.29.

Step 3: We can now state that the time factor is 64 calendar days. We can further state that our price factor is $4.29 minus $4.16 = $0.13.

Step 4: From the April 30 date extend a line so that it passes through the time co-ordinate (April 30 + 64 days = July 4) and the price co-ordinate $4.29. This line is the Gann 1x1 line.

Step 5: From the April 30 date, extend another line so that it passes through the time co-ordinate (April 30 + 2 x 64 days = September 8) and the price co-ordinate $4.29. This line is the Gann 1x2 line.

Step 6: From the April 30 date, extend another line so that it passes through the time co-ordinate (April 30 + (0.33) x 64 days = May 21) and the price co-ordinate $4.29. This line is the Gann 3x1 line.

The Wheat price chart in Figure 85 has these Gann lines overlaid starting from the April 30 low. From September 2019 until the wheat price peaked in early 2020, the 3 x 1 line provided guidance for the price trend. The 1 x 1 line is now underlying support.

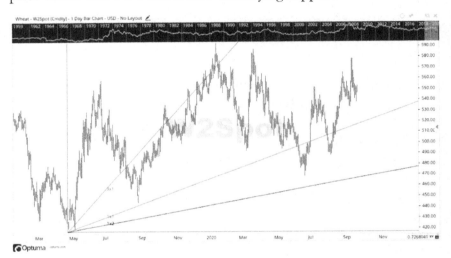

Figure 85
Gann Lines applied to a Wheat Chart

Price square Time

The concept of price square time says that at a significant swing point and trend change on a stock chart, commodity price chart or index chart exists because price and time have squared with one another. That is, a planet has advanced a certain number of degrees and price has changed by that same number of degrees or a multiple of those degrees.

To illustrate, Figure 86 presents a segment from a Gold futures chart dating to 2018-19.

142

Price Rise = $187

Figure 86
Gold price and the concept of Price Square Time

This chart shows that from August 18 through February 2019 there was a clear uptrend, which broke in February 2019. The price rise from low to high (intraday) was $187.

During this timeframe, Sun advanced 187 degrees (23 Leo to 0 Pisces) and Venus advanced from 8 Libra to 12 Capricorn (94 degrees ; 94 x 2 = 188), Mars moved from 29 Capricorn to 2 Taurus (93 degrees ; 93 x 2 = 186). Mercury and Jupiter moved 210 degrees and 35 degrees respectively, so they are not significant to this argument.

The planetary movement of Sun, Venus and Mars squares (closely matches) the price movement of Gold. Identifying price square time events will take some work on your part. But, the results will make the effort worthwhile. When dealing with large price values, such as the Dow Jones, the Nasdaq or the S&P 500, you will end up using a multiplier of the planetary movement values. To illustrate, consider the panic sell-off surrounding the COVID situation in early 2020.

From the February 12, 2020 trend turning point to the March 23 panic low point, the S&P 500 dropped 1149 points (close to close basis).

143

During this same time frame, heliocentric Venus moved 23 degrees. Take 50 times 23 and the result is 1150. Thus, one can argue that price and time had squared at the March 2020 lows.

During this same time frame, heliocentric Mars moved 16 degrees. Taking 72 times 16 gives a result of 1152. Thus, one can argue further argue that price and time had squared at the march lows.

I have yet to fully comprehend the mysteries of price squaring with time. Why in one instance does a 20 X multiplier work, when in the other instance a 72 X factor is needed? Why are these multipliers always whole integer numbers?

If you are prepared to take the time to determine both geocentric and heliocentric movements of Sun, Venus and Mars, price square time will prove to be a powerful tool for you.

Newton and Einstein

In the early 1700s and scientist Sir Isaac Newton developed his Theory of Universal Gravitation in which he said planets in our solar system are attracted to one another by gravity. Newton further said that space and time were absolute and that the world functioned according to an absolute order. Furthermore, he said that space was a three-dimensional entity and time was a two-dimensional entity.

In the early 1900's, Albert Einstein advanced his Theory of Relativity that posited Newton's absolute model was outdated. Einstein said the passage of time of an object was related to its speed with respect to that of another observer. Thus was penned the concept of relative space-time in which space was not uniform.

Einstein further stated that relative space-time could be distorted depending on the density of matter. That is, space-time in the area of the Sun is more distorted because the Sun has a great, huge mass. Light particles travelling near the Sun are then distorted from their linear path due to the mass of the Sun.

Quantum Price Lines

Quantum Price Lines are based on this quantum theory. The whole notion of Quantum Lines posits that the price of a stock, index or commodity can be thought of as a light particle or electron that can occupy different energy levels or orbital shells.

Author and market researcher Fabio Oreste combined the notion of quantum price lines with Einstein's theory that the fabric of space-time can be bent. Picture a group of people holding the edges of a large blanket. They pull on the edges until the blanket is stretched tight. Next, someone places a ball on the tight blanket. The weight of the ball causes a slight sag in the blanket fabric. Oreste says the point of maximum curvature of the sagged portion is akin to a Quantum Price Line. In his book book entitled, *Quantum Trading,* Oreste details his formula for Quantum Price Line calculation :

Quantum Line = (N x 360) + PSO ;

Where PSO = heliocentric planetary longitude x Conversion Scale
Where N is the harmonic level = 2^n ; 1,2,4,8,16 ....
Where Conversion Scale = 2^n ; 1,2,4,8,16 ....
When dealing with prices less than 360, the inverse variation of the formula is used.

Quantum Line = (1/N x 360) + PSO

The technique then allows one to calculate various sub-divisions of these Quantum Lines. Taking the value of the calculated Quantum Line, one would generate the sub-divisions by multiplying by 1.0625, 1.125, 1.875, 1.25, etc... in steps of 0.0625.

Please note the use of heliocentric planetary data in these Quantum Line calculations. There are websites that will provide you with this data such as www.astro.com/swisseph.

Alternatively, you can find a Heliocentric Ephemeris book such as *The American Heliocentric Ephemeris, 2001-2050.*

To assist you with calculating Quantum Lines, consider the following example:

On a given date, the following heliocentric planetary positions were noted: Mars 306 degrees, Jupiter 307 degrees, Neptune 324 degrees, Pluto 271 degrees.

In this example, let N=1 and let the conversion scale be set to CS=1.

The Oreste point of maximum curvature for these planets is then:

Mars: 360 + 306 = 666
Jupiter: 360 + 307 = 667
Neptune: 360 + 324 = 684
Pluto: 360 + 271 = 631

If you were to take another date in the future and calculate the points of maximum curvature, you could then join the two points for each planet. By definition two points joined equals a line. You could extend these lines out into the future. These lines are called Quantum Lines (or QL's).

If the above numbers seem oddly familiar, that is because they are. The S&P 500 March 6th, 2009 lows delivered an intra-day low of 665.7 and on the day the close was 687. Indeed. Mars, Jupiter and Neptune all acted in concert in March 6, 2009 to provide a floor of support under the US equity market.

Consider another example:

On a given date, the following heliocentric planetary positions were noted: Mars 355 degrees, Jupiter 296 degrees, Neptune 349 degrees, Pluto 293 degrees.

In this example, let's set the conversion scale as CS=2. Recall our formula is:
(N x 360) + (heliocentric position x conversion scale)
Calculating the back half of the formula gives us:

Mars: 2*355 = 710
Jupiter: 2*296 = 592
Neptune: 2*349 = 698
Pluto: 2*293 = 586

Add to these figures, a sum of 8 x 360 = 2880.
Mars: 710 + 2880 = 3590
Jupiter: 592 + 2880 = 3472
Neptune: 698 + 2880 = 3578
Pluto: 586 + 2880 = 3466

If these numbers seem familiar, so they should. The high on the S&P 500 on Sept 2nd, 2020 was 3579, which is the Neptune and arguably the Mars points of maximum curvature. These two characters are the "ruling planets" of the NYSE. In 2009, they helped define the March lows. Now they define the September 2020 highs. A co-incidence? I doubt it very much!

What follows is a suggested list of the Mars, Neptune and Pluto Quantum lines you may wish to overlay onto your various charts for 2021.

S&P 500 Index

Pluto, Neptune and Mars quantum lines (Conversion Scale =2, N=8) tend to work quite well for the S&P 500 Index. During 2021, consider drawing the following suite of Quantum Lines onto your daily chart of the S&P500 Index. Each line should start at January 1, 2021 and terminate at December 31, 2021.

Pluto - January 2021	Pluto - December 2021
3828	3832
3468	3472
3108	3112
2748	2752
2388	2392

Neptune - January 2021	Neptune - December 2021
3940	3944
3580	3584
3220	3224
2860	2864
2500	2504
2140	2144
3580	3584

Mars – January 2021	Mars - December 2021
3732	4070
3372	3710
3012	3350
2652	2990
2292	2630

Nasdaq Composite Index

Pluto, Neptune and Mars quantum lines (Conversion Scale=2, N=32) tend to work quite well for the Nasdaq Index. During 2021, consider drawing the following suite of Quantum Lines onto your daily chart of the Nasdaq Index. Each line should start at January 1, 2021 and terminate at December 31, 2021.

Pluto - January 2021	Pluto - December 2021
12108	12112
11748	11752
11388	11392
11028	11032
10668	10672
10308	10312
9948	9952
9588	9592
9228	9232
8868	8872
8508	8512
8148	8152
7788	7792
7428	7432

Mars - January 2021	Mars - December 2021
11652	11990
11292	11630
10932	11270
10572	10910
10212	10550
9852	10190
9492	9830
9132	9470
8772	9110
8412	8750
8052	8390
7692	8030

Neptune - January 2021	Neptune - December 2021
12220	12224
11860	11864
11500	11504
11140	11144
10780	10784
10420	10424
10060	10064
9700	9704
9340	9344
8980	8984
8620	8624
8260	8264
7900	7904
7540	7544

FTSE 100 Index

Pluto, Neptune, and Mars quantum lines (Conversion Scale =2, N=32) for the FTSE Index are shown in the tables following. During 2021, consider drawing the following suite of Quantum Lines onto your daily chart of the FTSE Index. Each line should start at January 1, 2021 and terminate at December 31, 2021.

Pluto - January 2021	Pluto - December 2021
8508	8512
8148	8152
7788	7792
7428	7432
7068	7072
6708	6712
6348	6352
5988	5992
5628	5632
5268	5272
4908	4912
4548	4552

Neptune- January 2021	Neptune - December 2021
8620	8624
8260	8264
7900	7904
7540	7544
7180	7184
6820	6824
6460	6464
6100	6104
5740	5744
5380	5384
5020	5024
4660	4664

Mars - January 2021	Mars - December 2021
8772	9110
8412	8750
8052	8390
7692	8030
7332	7670
6972	7310
6612	6950
6252	6590
5892	6230
5532	5870
5172	5510
4812	5150
4452	4790
4092	4430

S&P ASX 200 Index

Pluto, Neptune, and Mars quantum lines (Conversion Scale =2, N=32) for the ASX 200 Index are shown in the tables following. During 2021, consider drawing the following suite of Quantum Lines onto your daily chart of the ASX 200 Index. Each line should start at January 1, 2021 and terminate at December 31, 2021.

Pluto - January 2021	Pluto - December 2021
7788	7792
7428	7432
7068	7072
6708	6712
6348	6352
5988	5992
5628	5632
5268	5272
4908	4912
4548	4552
4188	4192
3828	3832
3468	3472

Mars - January 2021	Mars - December 2021
7692	8030
7332	7670
6972	7310
6612	6950
6252	6590
5892	6230
5532	5870
5172	5510
4812	5150
4452	4790
4092	4430
3732	4070
3372	3710

Neptune- January 2021	Neptune - December 2021
7900	7904
7540	7544
7180	7184
6820	6824
6460	6464
6100	6104
5740	5744
5380	5384
5020	5024
4660	4664
4300	4304
3940	3944
3580	3584

Gold Futures

Pluto, Neptune, Jupiter and Mars quantum lines for Gold are shown in the tables following. Each line should start at January 1, 2021 and terminate at December 31, 2021.

Pluto- January 2021	Pluto- December 2021
3468	3472
3108	3112
2748	2752
2388	2392
2028	2032
1668	1672
1308	1312
948	952

Neptune- January 2021	Neptune- December 2021
3580	3584
3220	3224
2860	2864
2500	2504
2140	2144
1780	1784
1420	1424
1060	1064

Mars- January 2021	Mars- December 2021
3012	3350
2652	2990
2292	2630
1932	2270
1572	1910
1212	1550
852	1190

Jupiter- January 2021	Jupiter - December 2021
3492	3552
3132	3192
2772	2832
2412	2472
2052	2112
1692	1752
1332	1392
972	1032

Silver Futures

Pluto, Neptune and Mars Quantum Lines (CS and N = 1/64, 1/32, and 1/16) are shown in the following tables. During 2021, consider drawing these Quantum Lines onto your daily chart of Silver. Each line should start at January 1, 2021 and terminate at December 31, 2021.

Pluto- January 2021	Pluto- December 2021
40.87	41
38.3	38.43
35.73	35.86
33.16	33.29
30.59	30.72
28.02	28.15
25.45	25.58
22.88	23.01
20.31	20.44
19.05	19.18
17.79	17.92
16.53	16.66
15.27	15.4
14.01	14.14
12.75	12.88
11.49	11.62
10.23	10.36

Neptune- January 2021	Neptune- December 2021
44.37	44.5
41.6	41.73
38.83	38.96
36.06	36.19
33.29	33.42
30.52	30.65
27.75	27.88
24.98	25.11
22.21	22.34
19.44	19.57
16.67	16.8
13.9	14.03
11.13	11.26
8.36	8.49

Mars- January 2021	Mars- December 2021
43.266	60.45
39.938	55.81
36.61	51.17
33.282	46.53
29.954	41.89
26.626	37.25
24.966	34.93
23.306	32.61
21.646	30.29
19.986	27.97
18.326	25.65
16.666	23.33
15.006	21.01
13.346	18.69
12.513	17.53
11.68	16.37
10.847	15.21

Currency Futures (Canadian Dollar, Australian Dollar, Japanese Yen)

Pluto, Neptune and Mars Quantum Lines (CS and N = 1/1024 and 1/512) are shown for these currencies in the following tables. During 2021, consider drawing these Quantum Lines onto your daily charts. Each line should start at January 1, 2021 and terminate at December 31, 2021.

Mars- January 2021	Mars- December 2021
0.8319	1.162
0.7799	1.0894
0.7279	1.0168
0.6759	0.9442
0.6239	0.8716
0.5719	0.799
0.5199	0.7264

Pluto- January 2021	Pluto- December 2021
1.277	1.281
1.1972	1.2009
1.1174	1.1208
1.0376	1.0407
0.9578	0.9606
0.878	0.8805
0.7982	0.8004
0.7184	0.7203
0.6386	0.6402

Neptune- January 2021	Neptune- December 2021
1.386	1.39
1.2994	1.3032
1.2128	1.2164
1.1262	1.1296
1.0396	1.0428
0.953	0.956
0.8664	0.8692
0.7798	0.7824
0.6932	0.6956

Currency Futures (Euro and British Pound)

Pluto and Neptune Quantum Lines (CS and N = 1/1024 and 1/512) are shown for these currencies in the following tables. During 2021, consider drawing these Quantum Lines onto your daily charts. Each line should start at January 1, 2021 and terminate at December 31, 2021.

Neptune- January 2021	Neptune- December 2021
1.386	1.39
1.2994	1.3032
1.2128	1.2164
1.1262	1.1296
1.0396	1.0428
0.953	0.956
0.8664	0.8692
0.7798	0.7824
0.6932	0.6956

Pluto- January 2021	Pluto- December 2021
1.277	1.281
1.1972	1.2009
1.1174	1.1208
1.0376	1.0407
0.9578	0.9606
0.878	0.8805
0.7982	0.8004
0.7184	0.7203
0.6386	0.6402

Wheat and Corn Futures

Pluto and Neptune Quantum Lines (CS and N = 1/256. 1/128 and 1/64) are shown for these grains in the following Tables. During 2021, consider drawing these Quantum Lines onto your daily charts. Each line should start at January 1, 2021 and terminate at December 31, 2021.

Pluto- January 2021	Pluto- December 2021
7.65	7.67
7.0125	7.03
6.375	6.39
5.7375	5.75
5.1	5.11
4.4625	4.47
3.825	3.83

Neptune- January 2021	Neptune- December 2021
8.308	8.328
7.616	7.635
6.924	6.942
6.232	6.249
5.54	5.556
4.848	4.863
4.156	4.17

Soybean Futures

Pluto, Neptune and Jupiter Quantum Lines (CS and N = 1/128 and 1/64) are shown for Soybeans in the following tables. During 2021, consider drawing these Quantum Lines onto your daily charts. Each line should start at January 1, 2021 and terminate at December 31, 2021.

Pluto- January 2021	Pluto- December 2021
10.2	10.23
9.56	9.59
8.92	8.95
8.28	8.31
7.65	7.67

Neptune- January 2021	Neptune- December 2021
11.07	11.1
10.38	10.40
9.69	9.71
9	9.02
8.30	8.32
7.61	7.63

Jupiter- January 2021	Jupiter- December 2021
10.38	10.85
9.73	10.17
9.08	9.49
8.43	8.81
7.79	8.14

Crude Oil Futures

Pluto, Neptune and Jupiter Quantum Lines (CS and N = 1/8 and 1/16) are shown for Crude Oil in the following tables. During 2021, consider drawing these Quantum Lines onto your daily charts. Each line should start at January 1, 2021 and terminate at December 31, 2021.

Pluto- January 2021	Pluto- December 2021
56.2	56.375
51.09	51.25
45.98	46.125
40.87	41
38.32	38.44
35.77	35.88
33.22	33.32
30.67	30.76
28.12	28.2

Neptune- January 2021	Neptune-December 2021
55.51	55.64
49.97	50.08
44.43	44.52
41.65	41.74
38.87	38.96
36.09	36.18
33.31	33.4
30.53	30.62
27.75	27.84
24.97	25.06

Jupiter- January 2021	Jupiter-December 2021
57.25	59.85
52.05	54.42
46.85	48.99
41.65	43.56
39.05	40.85
36.45	38.14
33.85	35.43
31.25	32.72
28.65	30.01
26.05	27.3

30-Year Bond Futures

Saturn Quantum Lines (CS and N = 1/8 and 1/16) are shown for 30-Year Bonds as follows. During 2021, consider drawing these Quantum Lines onto your daily charts. Each line should start at January 1, 2021 and terminate at December 31, 2021.

Saturn- January 2021	Saturn-December 2021
207.24	210.32
186.53	189.29
165.82	168.26
155.46	157.75
145.1	147.24
134.74	136.73
124.38	126.22

10-Year Treasury Note Futures

Saturn, Neptune and Pluto Quantum Lines (CS and N = 1/8 and 1/16) are shown for 10-Year Treasury Notes as follows. During 2021, consider drawing these Quantum Lines onto your daily charts. Each line should start at January 1, 2021 and terminate at December 31, 2021.

Saturn- January 2021	Saturn-December 2021
155.46	157.75
145.1	147.24
134.74	136.73
124.38	126.22

Neptune- January 2021	Neptune-December 2021
155.32	155.76
144.23	144.64
133.14	133.52
122.05	122.4
110.96	111.28

Pluto- January 2021	Pluto-December 2021
143.08	143.5
132.87	133.25
122.66	123
112.45	112.75
102.24	102.5

Sugar Futures

Pluto, and Neptune Quantum Lines (CS and N = 1/8 and 1/16) are shown for Sugar in the following tables. During 2021, consider drawing these Quantum Lines onto your daily charts. Each line should start at January 1, 2021 and terminate at December 31, 2021.

Pluto- January 2021	Pluto-December 2021
16.57	16.63
15.3	15.36
14.02	14.08
12.75	12.81
11.47	11.53
10.2	10.26
8.92	8.98

Neptune- January 2021	Neptune-December 2021
16.63	16.69
15.25	15.31
13.87	13.93
12.49	12.55
11.11	11.17
9.73	9.79

Cocoa Futures

Pluto, Neptune and Jupiter Quantum Lines (CS and N = 2, 4, 8) are shown for Cocoa in the following tables. During 2021, consider drawing these Quantum Lines onto your daily charts. Each line should start at January 1, 2021 and terminate at December 31, 2021.

Pluto- January 2021	Pluto-December 2021
3924	3936
3597	3608
3270	3280
2943	2952
2616	2624
2452	2460
2289	2296
2125	2132
1962	1968

Neptune- January 2021	Neptune-December 2021
3905	3916
3550	3560
3195	3204
2840	2848
2662	2670
2485	2492
2307	2314
2130	2136
1952	1958

Jupiter January 2021	Jupiter-December 2021
3996	4176
3663	3828
3330	3480
2997	3132
2664	2784
2497	2610
2331	2436
2164	2262
1998	2088
1831	1914

CHAPTER FOURTEEN

Conclusion

I have taken you on a wide ranging journey in this Almanac to acquaint you with the mathematical and astrological links between planetary activity and market price behavior. I sincerely hope you will embrace financial astrology as a valuable tool to assist you in your trading and investing activity. I hope you will pause often to contemplate whether the correlations you have learned about in this Almanac are the actions of the cosmos on the emotions of traders and investors or the actions of a select few power players using astrology to manipulate the markets.

If you decide to embrace financial astrology as a tool to help you navigate the markets, I encourage you to work hard and stick with it. At first it will all seem daunting. Fight the urge to give up. Soon enough it will all make sense and your trading and investing activity will take on a new meaning.

To encourage you to stick with it and master the use of astrology, I will leave you with the words of Neil Turok from his 2012 book, *The Universe Within*.

"Perseverance leads to enlightenment. And the truth is more beautiful than your wildest dreams".

GLOSSARY OF TERMS

Ascendant: one of four cardinal points on a horoscope, the Ascendant is situated in the East.

Aspect: the angular relationship between two planets measured in degrees.

Autumnal Equinox: (see Equinox) – that time of year when Sun is at 0 degrees Libra.

Conjunct: an angular relationship of 0 degrees between two planets.

Cosmo-biology: changes in human emotion caused by changes in cosmic energy.

Descendant: one of four cardinal points on a horoscope, the Descendant is situated in the West.

Ephemeris: a daily tabular compilation of planetary and lunar positions.

Equinox: an event occurring twice annually, an equinox event marks the time when the tilt of the Earth's axis is neither toward or away from the Sun.

First Trade chart: a zodiac chart depicting the positions of the planets at the time a company's stock or a commodity future commenced trading on a recognized financial exchange.

First Trade date: the date a stock or commodity futures contract first began trading on a recognized exchange.

Full Moon: from a vantage point situated on Earth, when the Moon is seen to be 180 degrees to the Sun.

Geocentric Astrology: that version of Astrology in which the vantage point for determining planetary aspects is the Earth.

Heliocentric Astrology: that version of Astrology in which the vantage point for determining planetary aspects is the Sun.

House: a 1/12th portion of the zodiac. Portions are not necessarily equal depending on the mathematical formula used to calculate the divisions.

Lunar Eclipse: a lunar eclipse occurs when the Sun, Earth, and Moon are aligned exactly, or very closely so, with the Earth in the middle. The Earth blocks the Sun's rays from striking the Moon.

Lunar Month: (see Synodic Month).

Lunation: (see New Moon).

Mid-Heaven: one of four cardinal points on a horoscope, the Mid-Heaven is situated in the South.

New Moon: from a vantage point situated on Earth, when the Moon is seen to be 0 degrees to the Sun.

North Node of Moon: the intersection points between the Moon's plane and Earth's ecliptic are termed the North and South nodes. Astrologers tend to focus on the North node and Ephemeris tables clearly list the zodiacal position of the North Node for each calendar day.

Orb: the amount of flexibility or tolerance given to an aspect.

Retrograde motion: the apparent backwards motion of a planet through the zodiac signs when viewed from a vantage point on Earth.

Sidereal Month: the Moon orbits Earth with a slightly elliptical pattern in approximately 27.3 days, relative to a fixed frame of reference.

Sidereal Orbital Period: the time required for a planet to make one full orbit of the Sun as viewed from a fixed vantage point on the Sun.

Siderograph: a mathematical equation developed by astrologer Donald Bradley in 1946 (By plotting the output of the equation against date, inflection points can be seen on the plotted curve. It is at these inflection points that human emotion is most apt to change resulting in a trend change on the Dow Jones or S&P 500 Index).

Solar Eclipse: a solar eclipse occurs when the Moon passes between the Sun and Earth and fully or partially blocks the Sun.

Solstice: occurring twice annually, a solstice event marks the time when the Sun reaches its highest or lowest altitude above the horizon at noon.

Synodic Month: during a sidereal month (see Sidereal Month), Earth will revolve part way around the Sun thus making the average apparent time between one New Moon and the next New Moon longer than the sidereal month at approximately 29.5 days. This 29.5 day time span is called a Synodic Month or sometimes a Lunar Month.

Synodic Orbital Period: the time required for a planet to make one full orbit of the Sun as viewed from a fixed vantage point on Earth.

Vernal Equinox: that time of the year when Sun is at 0 degrees Aries.

Zodiac: an imaginary band encircling the 360 degrees of the planetary system divided into twelve equal portions of 30 degrees each.

Zodiac Wheel: a circular image broken into 12 portions of 30 degrees each. Each portion represents a different astrological sign.

OTHER BOOKS BY THE AUTHOR

The Bull, The Bear and The Planets

Once maligned by many, the subject of financial astrology is now experiencing a revival as traders and investors seek deeper insight into the forces that move the financial markets.

The markets are a dynamic entity fueled by many factors, some of which we can easily comprehend, some of which are esoteric. *The Bull, The Bear and the Planets* introduces the reader to the notion that astrological phenomena can influence price action on financial markets and create trend changes across both short and longer term time horizons. From an introduction to the historical basics behind astrology through to an examination of lunar astrology and planetary aspects, the numerous illustrated examples in this book will introduce the reader the power of astrology and its impact on both equity markets and commodity futures markets.

The Lost Science

The financial markets are a reflection of the psychological emotions of traders and investors. These emotions ebb and flow in harmony with the forces of nature.

Scientific techniques and phenomena such as square root mathematics, the Golden Mean, the Golden Sequence, lunar events, planetary transits and planetary aspects have been used by civilizations dating as far back as the ancient Egyptians in order to comprehend the forces of nature.

The emotions of traders and investors can be seen to fluctuate in accordance with these forces of nature. Lunar events can be seen to align with trend changes on financial markets. Significant market cycles can be seen to align with planetary transits and aspects. Price patterns on stocks, commodity futures and market indices can be seen to conform to square root and Golden Mean mathematics.

In the early years of the 20th century the most successful traders on Wall Street, including the venerable W.D. Gann, used these scientific techniques and phenomena to profit from the markets. However, over the ensuing decades as technology has advanced, the science has been lost.

The Lost Science acquaints the reader with an extensive range of astrological and mathematical phenomena. From the Golden Mean and Fibonacci Sequence, to planetary transit lines and square roots through to an examination of lunar Astrology and planetary aspects, the numerous illustrated examples in this book will show the reader how these unique scientific phenomena impact the financial markets.

Stock Market Forecasting -
The McWhirter Method
De-Mystified

M.G. Bucholtz, B.Sc., MBA

Stock Market Forcasting: The McWhirter Method De-Mystified

Very little is known about Louise McWhirter, except that in 1937 she wrote the book *McWhirter Theory of Stock Market Forecasting*.

In my travels to places as far away as the British Library in London, England to research financial Astrology, not once did I come across any other books by her. Not once did I find any other book from her era that even mentioned her name. I find all of this to be deeply mysterious. Whoever she was, she wrote only one book. It is a powerful one that is as accurate today as it was back in 1937. The purpose of writing this book is suggested by the title itself – to de-mystify McWhirter's methodology.

THE COSMIC CLOCK
TIMING THE FINANCIAL
MARKETS USING
THE PLANETS

M.G. Bucholtz, B Sc, MBA

The Cosmic Clock

Can the movements of the Moon affect the stock market?

Are price swings on Crude Oil, Soybeans, the British pound and other financial instruments a reflection of planetary placements?

The answer to these questions is YES. Changes in price trends on the markets are in fact related to our changing emotions. Our emotions, in turn, are impacted by the changing events in our cosmos.

In the early part of the 20th century, many successful traders on Wall Street, including the venerable W.D. Gann and the mysterious Louise McWhirter, understood that emotion was linked to the forces of the cosmos. They used astrological events and esoteric mathematics to predict changes in price trend and to profit from the markets.

However, by the latter part of the 20th century, the investment community has become more comfortable in relying on academic financial theory and the opinions of colorful television media personalities, all wrapped up in a buy and hold mentality.

The Cosmic Clock has been written for traders and investors who are seeking to gain an understanding of the cosmic forces that influence emotion and the financial markets.

This book will acquaint you with an extensive range of astrological and mathematical phenomena. From the Golden Mean and Fibonacci Sequence through planetary transit lines, quantum lines, the McWhirter method, planetary conjunctions and market cycles. The numerous illustrated examples in this book will show you how these unique phenomena can deepen your understanding of the financial markets and make you a better trader and investor.

ABOUT THE AUTHOR

Malcolm Bucholtz, B.Sc., MBA, M.Sc. is a graduate of Queen's University (Faculty of Engineering) in Canada and Heriot Watt University in Scotland (where he received an MBA degree and a M.Sc. degree). After working in Canadian industry for far too many years, Malcolm followed his passion for the financial markets by becoming an Investment Advisor/Commodity Trading Advisor with an independent brokerage firm in western Canada. Today, he resides in Saskatchewan, Canada where he trades the financial markets using technical chart analysis, esoteric mathematics and the astrological principles outlined in this book.

Malcolm is the author of several books. His first book, *The Bull, the Bear and the Planets*, offers the reader an introduction to financial astrology and makes the case that there are esoteric and astrological phenomena that influence the financial markets. His second book, *The Lost Science*, takes the reader on a deeper journey into planetary events and unique mathematical phenomena that influence financial markets. His third book, *De-Mystifying the McWhirter Theory of Stock Market Forecasting* seeks to simplify and illustrate the McWhirter methodology. Malcolm has been writing the *Financial Astrology Almanac* each year since 2014.

Malcolm maintains a website (www.investingsuccess.ca) where he provides traders and investors with astrological insights into the financial markets. He also offers the *Astrology Letter* service where subscribers receive twice-monthly previews of pending astrological events that stand to influence markets.

RECOMMENDED READINGS

Astrology Really Works, edited by Jill Kramer for The Magi Society,(USA, 1995)

The Bull, the Bear and the Planets, M.G. Bucholtz, (USA, 2013)

The Lost Science, M.G. Bucholtz, (USA, 2013)

Stock Market Forecasting – The McWhirter Method De-Mystified, M.G. Bucholtz, (Canada, 2014)

The Cosmic Clock, M.G. Bucholtz (Canada, 2016)

The Universal Clock, J. Long, (USA, 1995)

McWhirter Theory of Stock Market Forecasting, L. McWhirter, (USA, 1938)

The Universe Within, N. Turok, (Canada, 2012)

A Theory of Continuous Planet Interaction, *NCGR Research Journal,* T.Waterfall, Volume 4, Spring 2014, pp 67-87.

Financial Astrology, Giacomo Albano, (U.K., 2011)

Made in the USA
Coppell, TX
19 January 2021

48421175R00105